Other books by Denise Carrington-Smith

Journeyings -
Along the Path with Edward Bach
From Rock-Rose to Rock-Water

The Enigma of Evolution
and the Challenge of Chance

Lord Lucan and Lady Luck -
The murder that never was

Outshining Darwin

Lamarck's Brilliant Idea

The true teaching of Lamarck explained

Storixus

DENISE CARRINGTON-SMITH

First published 2015 by Mossman Print

Second Edition published 2020 by
Storixus Independent Publishing
Canberra, Australia

www.storixus.com

ISBN 978-0-6483640-7-8 (Paperback Edition)
ISBN 978-0-6483640-8-5 (eBook Edition)

Author's Note

N O BOOK on evolutionary theory is complete without a mention, however brief, of the name of Lamarck. And brief most mentions are.

One thing they all seem to have in common – they all associate Lamarck's name with the discredited doctrine of the Inheritance of Acquired Characteristics, the belief that a change which occurs in the anatomy or physiology of a being during its lifetime can be passed on to its offspring.

When I pulled Lamarck's book from the shelf, I did so with the intention of enjoying some light and amusing holiday reading. I found myself entranced by the most profound book on evolutionary theory that I had ever read. Furthermore, I discovered that, far from endorsing the doctrine of the Inheritance of Acquired Characteristics, Lamarck specifically denied it!

I determined that there could be no better way of paying tribute to Lamarck than to mark the centenary of his book's translation in 1914 into my mother tongue than by bringing together a book which both cited and explained Lamarck's true teaching.

This book selects the principle points of Lamarck's teaching and quotes them in his own words. These words are then explained, a different type being used to avoid confusion.

The time has come for Lamarck to be reinstated in his proper place as the true founder of evolutionary theory.

Acknowledgments

I STUDIED Lamarck's theory while working on my Ph.D thesis at James Cook University, Cairns Campus. My supervisors, Drs. John Campbell and Liz (Elizabeth) Tynan encouraged me to write this book and I thank them for their support.

Otherwise, this has been a singularly solo effort – apart, of course, from Lamarck, whose spirit is always with me.

INSCRIPTION ON RELIEF

(Statue of Lamarck - Le Jardin des Plantes, Paris)

La postérité vous admirera,
elle vous vengera, mon père.

~

Posterity will admire you,
she will avenge you, my father.

THIS BOOK IS RESPECTFULLY DEDICATED

TO JEAN-BAPTISTE LAMARCK :

THE MAN AND THE MEMORY.

Content

Content

Section One

INTRODUCTION

THE year 2014 marked the one hundredth anniversary of the publication of an English translation of Lamarck's great work, *Philosophie Zoologique*. Lamarck's seminal work was the first published in the Western world devoted exclusively to the subject of zoological evolution. It was published in 1809, coincidentally the same year as that in which Charles Darwin was born. Lamarck's revolutionary ideas evoked much opposition and his work was in danger of being lost to the English speaking world, which fact inspired Hugh Elliot to undertake the laborious work of its translation.

In his *Introduction* to his translation, Elliot explained that, although many writers 'quoted' Lamarck, it would seem few had, in fact, read his work. They merely repeated the opinion of other writers before them. Of those who had indeed read Lamarck's work, some regarded him "as the greatest biological teacher that has ever lived". Elliot claimed that the outstanding feature of Lamarck's theory was its claim that species changed (evolved) over time, which ran counter to the prevalent theory of the stability of species.

The rapid expansion of building work following the industrial revolution, including larger roads and canals, which at times required the blasting away of massive amounts of material to allow their passage through, rather than over, a hill, had resulted in the exposure of many fossils, actual bones as well as impressions of both flora and fauna. Coal mining was another source of fossils, which were being collected and studied by 'antiquarians', rather than simply tossed aside, as had previously been the case. Later, the construction of railways would add greatly to the number and diversity of documented fossils. The great anatomist, George Cuvier (1769-1832), Lamarck's nemesis, believed that the Earth, when created but a few thousand years previously, had then been inhabited by a far greater number of species than now populated the Earth. Many of these species, claimed Cuvier, had been lost as a result of catastrophes, such as the Biblical flood, of which Cuvier suggested there had been several, that recorded in *Genesis* being the last. Other theorists postulated a series of 'creations'. Was there any reason to suppose that the Creator had exhausted his creative abilities with one act of Creation? Lamarck, alone, proposed a theory of gradual change, of evolution.

Both Buffon (see below) and Erasmus Darwin (Charles Darwin's grandfather) had mentioned the possibility of evolution, but their writings covered a wide range of subjects and neither followed through with their musings sufficiently to formulate anything which could be regarded as a theory of evolution. However, the topic was 'in the air' – and in the ground, as mentioned above, with the continual unearthing of new fossils. The English work which established evolution as an acceptable theory was that of Robert Chambers (1802-1861), who, in 1844, published his work, *Vestiges of the Natural History of Creation*. This book provoked much discussion and by 1846 had already run to five editions. The twelfth edition was published posthumously.

Robert Chambers and his brother ran a publishing company in Edinburgh, an acknowledged seat of learning due to the presence of Edinburgh's famous University. Their company specialized in books and pamphlets designed to inform the non-University class of reader about a wide variety of topics. They published the *Edinburgh Journal*, which was well respected and widely read. By 1842, Robert was suffering from exhaustion brought on by overwork and was forced to take two years off from work to recuperate. It was during this time that he wrote Vestiges.

In 1834, an English edition of Buffon's *Natural History, General and Particular containing the History and Theory of the Earth, and a General History of Man*, was published in London by Thomas Kelly. This edition included *A History of Fishes, Reptiles and Insects* by Henry Augustus Chambers, LL.D., possibly a relation of Robert Chambers. Be that as it may, it seems certain that Robert Chambers would have been aware of this translation of Buffon's work and it may well have ignited, if not merely fuelled, his interest in natural history in general and the possibility of evolution in particular. And so it happened that the year 1844, the centenary of Lamarck's birth, saw the publication in England of the work which firmly cemented evolution into the consciousness of the British reading public.

Today, Lamarck's name is associated with the discredited theory of the inheritance of acquired characteristics. This was not Lamarck's theory – it was Darwin's! Darwin called it *pangenesis* and, while it is 'assumed' in *On the Origin of Species*, published in 1859, Darwin fully outlined his theory in *The Descent of Man*, which was published in 1871. Quite how and why these theories became reversed, how Lamarck's name came to be associated with the inheritance of acquired characteristics and Darwin's with the theory of evolution, is a story for another day.

Quite why Elliot undertook his task of translation at the time that he did must forever remain a matter of speculation. I cannot but wonder whether he had been inspired by the fact that, five years previously in the year 1909, the Western world had celebrated the centenary of Darwin's birth? Books had been written (for example Bateson 1909), articles had been published, but none mentioned the name of Lamarck! It was the same one hundred years later, when the bi-centenary of Darwin's birth, and the 150th anniversary of the publication of Darwin's work, *On the Origin of Species by Means of Natural Selection*, had similarly been celebrated, without mention of the fact that 2009 was also the bi-centenary of the publication of Lamarck's ground-breaking work. There was sufficient interest in Lamarck's work in the English speaking world for a second edition of Elliot's translation to be published in 1963. Nevertheless, Lamarck's work continued to be largely ignored, or, if mentioned, misrepresented. Part of the responsibility for that, I believe, must be laid at the feet of Hugh Elliot. Elliot did not merely translate Lamarck's work, he provided a lengthy *Introduction* in which he outlined his understanding of Lamarck's theory. I believe this interpretation to be seriously flawed. Before giving my understanding of Lamarck's theory, it is time to answer the question: Who was Lamarck, anyway?

Jean-Baptiste-Pierre-Antoine de Monet Chevalier de Lamarck (1744-1829)

JEAN-BAPTISTE Lamarck was born in France, at a place called Bazantin in Picardy, on 1st August, 1744, the youngest of eleven children. He came from an old established family, his father, Philippe Jacques de Monet, being lord of a manor, but they were of limited means, not members of the wealthy French aristocracy. Lamarck married late. In 1777, at the age of 32 or 33, he formed a lasting relationship with Marie-Françoise de Fontaines de Chuignolles, who bore him six children. However, they were not married until 1792, the ceremony being performed when Marie-Françoise was on her deathbed. While it was not uncommon for peasant people to live in *de facto* relationships, for a person of Lamarck's standing, the following of this path must have been a deliberate decision. Perhaps his antipathy towards the Church (see below) discouraged him from entering its portals and his principles prevented him from participating in a ceremony endorsed by a Church in which he did not believe. As will soon become apparent, Lamarck early showed an independent streak

4

and all his life followed his own direction, popular or not.

Having taken the plunge once, Lamarck must have found the experience not as traumatic as he had anticipated, for he married a further three times, fathering two more children, although the legitimacy of his final relationship is not certain.

His three older brothers having made careers in the Army, it was his father's decision that Jean-Baptiste should enter the Church and he was placed in the Jesuit College at Amiens in 1755, while still only 11 years old. He was to spend four years there. His father died in 1759 and Jean-Baptiste took the opportunity to make a career change. He ran away to join the Army! At that time, the French were fighting in Germany, it being close to the end of the Seven Years' War. He purchased a horse, acquired a letter of introduction from a friend and, thus equipped, joined the French Grenadiers on the eve of the Battle of Fissinghausen. He was then 16 years of age. The French were soundly defeated, the officers of Lamarck's company being killed. It is reported that he took charge and showed courage under fire, which resulted in his immediate appointment as an officer. In a footnote, Elliot questioned the accuracy of this account, which was based upon a letter written to Cuvier in 1830, shortly after Lamarck's death, by one of his sons. Elliot felt that the letter magnified Lamarck's achievements. However, I find nothing untoward in the account of this part of his life. Teenage boys are notoriously hot-headed and feel themselves to be invincible. Lamarck's family had a long history of military involvement and his three older brothers had been allowed to follow this path. That a boy of his age should prefer a military, rather than a Church, career is quite understandable. The officers being killed, it would have been imperative that a substitute to be appointed with all speed. In those days, officers did not work their way up through the ranks; they were appointed from among the ruling classes, not merely because of an assumed ability to lead, but because they were educated. An ability to read and write far greater than that acquired by most peasant children at the local village school, was essential. The arrival of a young man thus qualified must have been seen as very fortuitous by the Field Marshall who commissioned him. Today, we think of 16 year-olds as school children. Such was not the case then when children entered the work force, took apprenticeships or entered their vocational training at a College or University, away from home, by approximately twelve years of age. Lamarck had spent four years at College before his 'escape' and would have been considered a man.

Lamarck was discharged from military service on medical grounds at the still young age of 22. He suffered an enlargement of the cervical glands which condition, it was suggested, had resulted from 'horseplay' in the barracks, during which Lamarck had been severely pulled by his hair, stretching his neck. By way of footnote, Elliot gave a caveat, explaining that this was the account given by Cuvier, which differed from that given by his son in the above cited letter. Although Elliot had seen the letter, he gave no further details. While such 'horseplay' may have aggravated a pre-existing condition, it is unlikely that it was the primary cause of an enlargement sufficiently severe to require an operation and subsequent discharge from the army. Whatever the condition from which Lamarck suffered, it either resolved itself, or the operation performed was successful, because no mention is made of any further related problem during the remaining sixty-three years of his life.

Lamarck's life's work

LAMARCK had now to decide upon a further career. The Church, the Army, Medicine or Law were the acceptable paths open to younger siblings of the aristocracy. Lamarck had already made an 'escape' from the Church and his chosen career in the Military was now closed to him. With his capacity for detailed and logical thought, it may be supposed that Lamarck would have been well suited to the Law, but he chose Medicine instead. He moved to

Paris, where he lived in a top floor apartment (garret) for a short while, before moving in with his brother. For a year, he supported himself by working as a bank clerk; then he took up his medical studies, which lasted four years (1767-1771). At this time, medical treatment was largely based on the prescription of herbal remedies, although the use of chemicals, such as arsenic and mercury, was becoming increasingly popular. Botany formed a major part of Lamarck's study and it was in this discipline he was to specialize, never practicing as a doctor. The *Jardin du Roi* (later known as *Le Jardin des Plantes*) was not merely a botanical garden where one could view plants from around the world, as were the Royal Botanical Gardens at Kew in England. Because of the close connection between herbs and medicine, the botanical gardens in Paris were also the centre for medical education and biological research.

It was during this time that Lamarck made the acquaintance of George Louis Leclerc, Comte de Buffon (1707-1788), usually referred to simply as 'Buffon'. Buffon was an eminent natural philosopher, who had regularly published a series of papers on subjects ranging from the stars and galaxies to the structure of the Earth, its plants and its animals. These came together as *Natural History, General and Particular, containing the History and Theory of the Earth* (Buffon 1781), forming a sizable volume of work for which Buffon was rightly held in high esteem. Buffon became a patron of Lamarck. After ten years' study and work, Lamarck, with the assistance of Buffon, published *Flora Française* (1781), a comprehensive account of the flowering plants of France. This resulted in Lamarck being admitted to the French Academy of Science.

Buffon assisted Lamarck in other ways. On his recommendation, Lamarck was appointed 'Botanist to the King'. The education of both Lamarck and of Buffon's son was extended when Lamarck was chosen by Buffon to accompany his son on a two year tour of Germany, Hungary and Holland (1781-1782), where they studied rare plants and had the opportunity of meeting other eminent botanists. On his return, Lamarck was appointed keeper of the Herbarium at the *Jardin*, writing his *Dictionaire de Botanique* and *Illustrations de Genres*. After Buffon's death in 1788, Lamarck continued his work at the *Jardin*. He recommended its re-organization, submitting a proposal to the *Assemblée Nationale* which was accepted, in large part, when, in 1793, the *Jardin* became the *Museum d'histoire Naturelle*. The new complex was extended to cover twelve different scientific fields, each overseen by its own professor. No longer was it merely a 'Garden'. It now encompassed the study of animals. There were two chairs of zoology and Lamarck was appointed Professor of the department dealing with 'insects and worms', as Linnæus had termed them. The Chair for 'superior' animals (mammals, birds, reptiles and fish) was awarded to Geoffroy St. Hilaire.

It might have been supposed that the eminent Lamarck would have been appointed to the chair of botany. This was not the case. In his *Introduction* to his translation of Lamarck's *Philosophie Zoologique*, Hugh Ellot gave the following account of events:

> Two chairs of zoology were created: one of which was devoted to mammals, birds, reptiles and fishes, while the other was devoted to the "inferior animals" (the insects and worms of Linnaeus), or, as Michelet called it, "l'inconnu". To the first chair, Geoffroy Saint-Hilaire was appointed, then a young man of 22. For the second chair, containing the unknown part of the Animal Kingdom, there were no obviously suitable candidates. Lamarck was a botanist of 25 years standing, but the chair of botany had passed to Desfontaine, and there now seemed nothing suitable remaining for him except this chair of zoological remnants.

This account gives the impression of a man rejected for promotion, not because of his incompetence, but because of his age. Is there another possible explanation for Lamarck's appointment to the 'inferior' Chair? I believe there is. Is it not at least possible that, during his years of botanical study, Lamarck became genuinely interested in the beetles and bugs, butterflies and bees, as well as all the other many, varied and incredible species with which our plants have such a close relationship? Is it not at least possible that Lamarck, after having devoted more than twenty years to the study of plants, requested a change, a chance to break new ground? As Elliot pointed out, the 'inferior' animals were little known (inconnu) compared with the 'superior' animals, yet their anatomy, physiology, method of reproduction and overall way of life was so much more varied and interesting than that of the 'superior' animals, now known as vertebrates. These latter are distinguished by being carnivores, herbivores or omnivores and either egg-laying, live-bearing of young hatched from eggs while still in their mother's body, or live-bearing after development in the uterus and suckled, i.e. mammals. This was already known. Which was the greater challenge? That Lamarck may have chosen the invertebrates must remain speculation, but so, too, must the idea that he was allocated their study as some form of 'consolation prize'.

Honeywill (2008) supported Elliot's claim that the Chair for the study of inferior animals had been awarded to Lamarck because it was not wanted by anybody else. However, Honeywill added that Lamarck later purchased a modest summer house with the proceeds of the sale to the Government of 'one of his spectacular shell collections'. If the sale of one of Lamarck's 'spectacular shell collections' was sufficient to purchase a house, however modest, what does that say of the extent and value of Lamarck's complete collection? Does not Lamarck's obvious interest in shells indicate that he already had a deep interest in at least some invertebrates prior to being awarded the Chair? It was Lamarck who had recommended that the Jardin should be, not merely reorganized, but extended to encompass animals as well as plants. It was he who wrote the recommendations which were submitted to the Assemblée in 1793. He was at that time held in high regard on account of his botanical work, his controversial ideas on evolution not yet having been enunciated, let alone published. Why should he not have been awarded a Chair, acceptable to him, on the grounds of his work? He was to receive a salary of 2800 francs per annum. Overall, his promotion was considerable. One of the first things which Lamarck did, after taking up his new position, was introduce the terms 'vertebrate' and 'invertebrate' to replace the terms 'superior' and 'inferior'. There is nothing in Lamarck's work to suggest that he considered any life form to be 'superior' or 'inferior' to any other, including humans, merely different, more or less complex.

There was great excitement in Paris in 1804 when Captain Nicolas Baudin returned from an expedition to the southern hemisphere which had taken him around Australia. This occurred at the same time as Captain Matthew Flinders was making a similar, competing expedition on behalf of the English. The French and the English were then at war, but Baudin and Flinders brokered an unofficial amnesty, which allowed the meeting of their respective expeditions at Encounter Bay, South Australia, to be amicable. It was a battle, not of guns, but to acquire the best collection of fauna and flora and a race to be the first to return home with their prizes. Baudin won, in part because, when Flinders called in at the French island of Mauritious for repairs, the French Governor arrested him as a spy and held him prisoner for six years! When Baudin arrived back in Paris in 1804, the Empress Josephine laid claim to the emus and the kangaroos, which she kept in her extensive gardens. The thought of travelling for months in what was by today's standards quite a small ship, with emus and kangaroos on board, is mind-boggling, but they did! Thus Lamarck would have had access to extraordinary specimens of creatures previously unknown to the northern hemisphere, other than possibly by drawings, some alive, others preserved, but at least 'in the flesh'. This occurred five years before Lamarck published his Philosophy Zoologique, although, of course, he was constantly publishing other material.

Lamarck lived in turbulent times. As a young man, he grew up during a time when a King ruled over France. When he fought in the Army, he did so in the service of the King. He relocated to

Paris in time to experience the trauma of the French Revolution, an event without precedence in European history. The deposition, trial and execution of the English King, Charles I, more than a century before, had been carried out with a semblance of dignity and legality. Although this event was followed by civil war before the Restoration which saw Charles II reclaim the throne, there was nothing which happened in England which could be compared with the blood-letting that characterized the French Revolution. Lamarck was from a minor aristocratic family, albeit the youngest of eleven children. Nevertheless, to live, work and survive in Paris during this time was an achievement in itself. Lamarck does seem to have been a genuine supporter of Napoleon, possibly because Napoleon overthrew the Catholic Church. Napoleon permitted a belief in a 'Supreme Being' but would not tolerate Christianity, showing his contempt by using Nôtre Dâme cathedral to stable his horses! Remembering that the young Lamarck had 'escaped' from the Jesuits at the first opportunity, it is reasonable to assume that he genuinely embraced Napoleon's vision.

The acme of Lamarck's career coincided with the acme of Napoleon's power. Part of Lamarck's duties at the Jardin was the giving of lectures. It was during this time, the 1790s, that Lamarck first formulated his ideas on evolution, publishing some early papers. In 1809, he published the ideas he had been formulating as the result of his work. His *Philosophie Zoologique* outlined his theory of evolution and was the first book ever to be published devoted solely to this topic. Lamarck's later work, *Histoire naturelle des Animeaux sans vertèbres*, was published in a number of volumes between 1815 and 1822. By that time Napoleon had been defeated at the Battle of Waterloo. As Napoleon's fortunes faded, so did those of Lamarck. The Catholic Church became re-established and his younger colleague, Georges Cuvier (1769-1832), a devout Catholic, completely rejected any theory of evolution which could not be reconciled with the Biblical account of creation and the establishment of Man on Earth as told in the book of *Genesis*. Cuvier, who had taken over St. Hilaire's position at the Museum d'Histoire Naturelle when St. Hilaire accompanied Napoleon to Egypt, became known as a brilliant anatomist. It was said that he could reconstruct a complete animal from one bone, so great was his understanding of the inter-relationship of the bones of the body. Cuvier was the first person to subject fossil bones to scientific scrutiny and to attempt to classify them into families and species. Cuvier acknowledged that there were fossil remains of creatures which no longer existed and that these may have become extinct as a result of a natural catastrophe, such as a flood. He even acknowledged that the Flood recorded in *Genesis* may have been the last of a series, but he completely rejected the concept of change through the process of evolution. The high regard in which Cuvier was held, the falling out of favour of ideas associated with the Napoleonic regime, as well as the re-establishment of Christian views, all combined to counter Lamarck's revolutionary concept, which was rejected and derided. Cuvier campaigned against Lamarck's evolutionary ideas with a religious fervor – and won! Lamarck was discredited during his lifetime and Cuvier's attacks continued after his death. How different the history of evolutionary theory may have been had Geoffroy St. Hilaire not surrendered his Chair to Cuvier when he chose to accompany Napoleon. According to Honeywill, St. Hilaire and Lamarck were like-minded; St. Hilaire supported Lamarck's ideas and the two men were life-long friends.

Unfortunately, although Lamarck supported Napoleon, he does not seem to have received reciprocal support in return. Once again, allow Elliot (1914: xxi) to tell the tale:

> The scientific world of his time rejected his [Lamarck's] theories of transformation; Cuvier, who was firmly convinced of the fixity of species, became the most famous and fashionable biologist of the time, and Lamarck's influence was completely overshadowed. Arago, in his Histoire de ma Jeunesse, relates the story of his meeting with Napoleon. The Emperor was receiving the Members of the Institute at the Tuileries, and

Lamarck attended, carrying with him his latest work, which happened to be the Philosophie Zoologique, to present to Napoleon. Napoleon first spoke to Arago, who had just been elected to the Institute, and then passed to Lamarck. "Napoleon," says Arago, "passed from me to another member of the Institute: a naturalist famous for his brilliant and important discoveries, M. Lamarck. The old man presented Napoleon with a book. "What is this?" said the Emperor. "Is it your absurd Météorologie with which you are disgracing your old age? Write on natural history, and I will receive your works with pleasure. This volume I only accept out of consideration for your grey hair. Here!" and he handed the book to an aide-de-camp. Lamarck, who had been vainly endeavouring to explain that it was a work on natural history, was weak enough to burst into tears.

I find this anecdote difficult to accept as given. That Napoleon could have made a derisive comment about Lamarck's almanacs in the privacy of his chambers, and that one of his servants had been indiscrete enough to repeat his remark, I can believe. However, courtly manners were as valued in France as they were in England. France may no longer have a King, but she did have an Emperor, who would have been very aware of his public image. This Emperor was greeting guests at a formal reception. I cannot believe that he would have publically humiliated one of his guests, particularly one "famous" for his "brilliant and important discoveries". Again, I can believe that Napoleon received the book but briefly before handing it to his aide. That would be normal practice. I can also believe that Napoleon may have cut short any attempt of Lamarck to enter into an explanation of his latest work – Napoleon had other guests to greet and this was not the appropriate time for engaging in extended conversation. That Lamarck burst into tears in public, I doubt, although he may have been distressed. After all, he had been working on his theories for nearly twenty years; it was the equivalent of two or three Ph.Ds. To have it seemingly set aside would be 'enough to make a grown man cry'! However, I suspect this incident has been embroidered as part of the campaign to discredit Lamarck and his work, which took place both during the latter part of his life and after his death. The incident, as reported by Arago and Elliot, does not sit well with what we know of French manners at this time.

There is a further reason why it is unlikely that Napoleon mistook Lamarck's new book for one of his meteorological almanacs. During the time he was living in the garret in Paris, Lamarck apparently spent much time watching the clouds, their formation, flow and dissolution. He became interested in the new science of meteorology and later tried his hand at long range weather forecasting. It is not surprising that he was unsuccessful. Only in the last few years have the meteorologists, with the aid of satellite technology, been able accurately to forecast the weather for about one week ahead. Predictions for further into the future are made with caution. Lamarck was brave, or foolish, enough to try publishing an annual meteorological almanac. His *Annuaires Métérologiques* was published from 1800-1810. It is this product which I have no difficulty in believing Napoleon may have disparaged. However, it seems unlikely that Napoleon would have mistaken one of these almanacs for the leather bound, three-part, nine hundred page two volume work which Lamarck presented to him. Lamarck's book took up so many pages because there was a limited amount of text on each page. It was fashionable at that time to provide extensive margins to allow the reader/student to make notes. Elliot's translation ran to just over four hundred pages.

Elliot further disparaged Lamarck when he referred to him, on his appointment at the age of 38 to the position of keeper of the Herbarium at the Jardin du Roi, as receiving a 'wretched' salary of 1000 francs a year (Honeywill used the term 'paltry'). It is difficult to make a direct comparison between the salaries of yesteryear and those of today. However, Lamarck received a pension of 400 francs per year upon his discharge from the army. His salary was two-and-a-half times that of his pension. A similar comparison between pension rates and salaries today would place 1000 francs as 'average' for a person of that age. However, today's average is inflated due to the very high salaries paid to CEO's, bank executives, industrialists, sport and T.V. stars, etc. In Lamarck's time, wealth was usually inherited, although some industrialists, such as those in the cotton industry in England, were making fortunes. However, neither they, nor the industrialists of the following century, were salaried. There were no persons receiving mega wages or salaries in the eighteenth century. A more appropriate comparison with today would be with the 'mean' or 'median' wage, not the average. Lamarck would have expected, and received, a 'comfortable' salary. Elliot cited Lamarck's salary in 1794, when he was appointed to the Chair of 'zoological remnants' at 2868 livres, 6 sous, 8 deniers, which would translate into English as £2868. 6s. 8d, a considerable sum by English standards. The *Jardin du Roi* was a prestigious place of employment and his pay would have been commensurate with that of his colleagues both then, and later during his time as Professor.

We should remember that at this time Lamarck was a renowned botanist. The controversy which was to accompany his later work, in which he disputed the stability of species in favour of a theory of evolution, was not yet conceived or perceived. There is no reason why his employers or his colleagues should demean his status, particularly bearing in mind that he was the protégé of the revered Count Buffon. Lamarck did have eight children to support, one of whom was deaf and another retarded. There is no doubt that Lamarck suffered financial hardship in his later years. Elliot's account does not give any information regarding Lamarck's later income. It is not clear whether Lamarck continued to receive a pension from the Army, possibly not, since he had been gainfully employed for so many years. Honeywill (2008) mentioned a pension of 1200 francs having been awarded to Lamarck in addition to his annual salary, but this seems to have been awarded around 1797, soon after he was made Professor, and may not have continued during the later, turbulent years, following the fall of Napoleon. Lamarck's sight started to fail him while he was yet in his sixties and he was blind at the end of his life. This was not an uncommon fate for older people in the days before cataract surgery. Undoubtedly, as he aged, he would have spent an increasing proportion of his financial reserves on medical treatment. He died on 18th December, 1829, and received a pauper's burial in a common burial ground, which was cleared every five years to receive further internments. The final resting place of his bones will never be known.

Elliot's role

HUGH Elliot took much trouble over the task which he had set himself. A translation such as this, even of a work only a hundred years old, entailed investigation into possible changes in the meaning of words and phrases, particularly those associated with botany and biology. Lamarck had often used the French, rather than the Latin, name for a species:

> In order to find out what animals he meant by these names, I have in the case of invertebrates referred to each one in Lamarck's later work, Animaux sans Vertèbres, second edition, 11 vols., where the French name is almost always given in conjunction with the Latin name (p.lvii).

Elliot continued to give other examples of texts which he had consulted in order to preserve the integrity of his translation and we are indebted to him for the care he took. Elliot found himself faced with a difficulty. Lamarck's work was 'ponderous' and 'tedious', consisting of long sentences with numerous subordinate clauses. Should he improve the readability of this work by making it more concise? Elliot decided against this course of action, instead determining to preserve the text in its original form, but to "write by way of introduction a brief précis of the whole work, stating as far as possible the sum of Lamarck's doctrine in my own words".

16

Most of Elliot's Introduction is scathing of Lamarck's work, leaving the reader wondering why Elliot took the trouble that he did to translate and preserve it. Strangely, it was the concluding section of Elliot's Introduction which was the most positive. He commenced by including Lamarck among "those great leaders of thought and action, who at one time exercise profound influence over their generation" pointing out that Lamarck defended the doctrine of organic evolution against the dominant thinking of leaders of both the Church and Science: "For almost half a century his writings stood as almost the only public representation of a belief which now no one questions. Then came the *Origin of Species*: a work which naturally and immediately superseded every earlier publication ...". In fact, it was less than a quarter of a century before the possibility of evolution was openly being canvassed, for example by persons such as Edward Blyth who first mooted the possibility of evolution by natural selection in the mid 1830s. The dominance of Darwin's theory after its publication may have been 'immediate' but there was nothing 'natural' about its selection, which was carefully crafted by his friends and supporters, principally Sir Charles Lyell and Thomas Huxley, who promoted Darwin's ideas for philosophical and political reasons. However, that, too, is a story for another time!

Elliot went on to write: "The scene has now changed once more: the reaction has in various quarters turned against Darwin, while Lamarck himself is slowly entering upon the final stage of oblivion." Some readers may not be aware that towards the close of the 19th century, Darwin's ideas fell into disfavour largely because Darwin and his supporters had been unable satisfactorily to explain how a new characteristic could spread through a population. Breeders of both plants and animals knew that unless their breeding stock was segregated from the general population, the desired characteristic would become diluted with each generation, eventually being 'blended out'. It was the rediscovery of Mendel's work in 1900, with its suggestion of 'discrete' inheritance whereby characteristics could be preserved,

17

unexpressed, on recessive genes, which saved Darwin's theory. By the time Elliot was undertaking his translation, Mendel's ideas were becoming increasingly well known but it was not until after the First World War that genetics became established as a separate scientific discipline and the melding of Darwin's and Mendel's theories became known as Neo-Darwinism.

Elliot considered Lamarck a 'second rank' philosopher because he appeared to have been "agnostic by reason and a deist by desire". Elliot was a materialist, denying the existence not only of the soul, but also of "an entity called mind" (page lxxxii) and found Lamarck's comment that he "liked to believe in a first cause or in short a supreme power which brought nature into existence" and "the universe as having some goal or purpose" (pp.lxxxviii-lxxxix), unsatisfactory. As a 'pure zoologist', Elliot considered that Lamarck had not the reputation of Cuvier, although "his judgment and method were of a very high order ... there can be no justification for the contempt with which many people now speak of Lamarck" (p.lxxxix). Elliot affirmed that Lamarck "never exploited science for his own advantage" but "abandoned his reputation in his immovable resolve to find the truth" (p.xci).

> Had he (Lamarck) been a soldier and suffered thus for the sake of the country, how great would have been the honour that would have rewarded so deep a devotion! But he was the soldier of no country: he was the soldier of humanity and truth alone. To my pen falls the lot of vindicating the memory of one who, if he had labored to destroy his fellow-men instead of to enlighten them, would have received all the glories of a national hero (p.xcii).

Misunderstandings

The most extreme example of what I believe to have been Elliot's misunderstanding of Lamarck's theory occurred on pages xxxiv-xxxv of his *Introduction*. This section commenced (page xxxiv):

> Lamarck held, then, that if it were not for the effects of
> environmental influences, the innate tendency to develop
> would be the exclusive factor in operation.

Lamarck held that there was an energy, which he called the *aura vitalis*, which caused it to strive to develop, to stretch the boundaries which contained it, to produce further forms of life.

In this his doctrine was similar to that of the Jesuit priest, Teilhard de Chardin, whose philosophical works were published posthumously in the 1950s, their publication having been forbidden during his lifetime by the Catholic Church. Elliot then continued (page xxxiv):

> We should then see the linear series of animals to be a
> perfectly regular and even progress in complexity of
> organization from to man. Each animal born would
> presumably be slightly more complex than its parent.

With no external influence from the environment deflecting its course, evolution would proceed in an orderly manner, i.e. in a straight line. So far, so good. But then (page xxxiv):

> If we could trace the ancestry of man, we should find as
> we went backwards that each individual was to an
> excessively small degree less complex than its immediate
> neighbour, till we finally ended with the infusorians.

At this point, Elliot is assuming that humans would have evolved into the form which they have today even without any influence from the environment. Lamarck never made any such supposition. Lamarck saw change, evolution, as occurring as the living form, be it infusoria, plant or animal, reacted to changing external conditions. Without this external stimulus, there would be no change; the form evolved would continue in its (then) current form. Having made this deviation from Lamarck's thinking, Elliot continued (page xxxiv-xxxv):

> All existing animals are on the road of development from
> Monas to man, and man's ancestors include every

existing species of animal. Not only had he bird, reptile and fish ancestors, but also arachnid, insect, worm, starfish, etc., ancestors. He passed through the stage of being a scorpion and a spider. He traversed in turn every known species of insect. He was a tapeworm, a sea-anemone, a polyp and an amoeba.

It is hardly surprising that Elliot deemed this doctrine to be "totally absurd". It is difficult to understand how a person who had so painstakingly translated Lamarck's work could so disastrously misunderstand it, let alone wish to preserve it!

The purpose of this book is to give a different explanation, one which I believe to be more true to Lamarck's ideas.

If it was Elliot's aim to encourage more people to read Lamarck's work, then, sadly, it would seem that he failed. Today, most texts contain but a short account of Lamarck's ideas, most of which appear to be but a regurgitation of an account given in a previous work. There are less than a handful of people who appear to have read Lamarck's book, who make comments along the lines: "Lamarck did not actually say what people think he said, but it is too late to worry about that now. The term 'Lamarckism' is associated with the Inheritance of Acquired Characteristics and to attempt to change that now would cause confusion. It is better to let things stay as they are".

In many ways, statements such as these are more disturbing than imperfect accounts penned in ignorance. Since when has science been served by the deliberate preservation of an acknowledged untruth? This 'Conspiracy of Silence' has resulted in the name of one of the Western world's greatest thinkers being vilified and disparaged.

I have accepted Elliot's translation as accurate, but not his explanation, which appears to me completely to misunderstand and misrepresent Lamarck's ideas. It must not be thought that, because I disagree with Elliot on some points, that I dismiss his opinion in its entirety. Nothing could be further from the truth.

However, it is not the purpose of this book to draw attention to those parts of his Introduction where I believe Elliot has correctly interpreted Lamarck's work. It is the purpose in this book to draw attention to those places where I believe Elliott, and others, have misunderstood Lamarck and to provide a commentary which I hope will make Lamarck's thinking more clear to the modern reader – and more acceptable.

While Eastern thought had always presented time and space as immeasurable, Western thinking had been far more limited and limiting. It took Galileo to break the fetters which bound Western thinking and allow the Western scientist to begin to grasp the enormity of space. What Galileo was to space, Lamarck was to time. It was Lamarck who first burst through the barrier of time, who first appreciated some of its immeasurable depth. Bear this in mind as you read his work and try to imagine what it must have been like for one man, alone and unsupported, to contemplate this grandeur.

Lamarck was one of the greatest thinkers the Western world has ever produced. He deserves to be remembered.

PREFACE

THE time has now come to look at Lamarck's work, starting with his Preface, proceeding from there to his Preliminary Discourse and thence to the individual chapters which comprise the three main parts of his thesis. Some reference will be made to other writers, such as Charles Darwin and Teilhard de Chardin, for the purpose of comparison, but the main emphasis from here on will be the work of Lamarck himself.

[Note: Where the present *font* is used, then the writer was me. Where this *font* is used, then the writer was Lamack. It may seem complicated but I think you will soon find yourself in the swing of things!]

In order to appreciate the depth, breadth and significance of Lamarck's thought, it will be imperative to consider it within its historical context. Reference has already been made to the turbulent political times through which Lamarck lived but consideration will also need to be given to the scientific thought of his time. Elliot (page xxi) commented on Lamarck's interest in chemistry. Apparently, Lamarck "attacked the chemistry of Lavoisier and Berthollet, which further completed the discredit in which his excursions outside biology involved him". I am quite happy to accept Elliot's opinion that chemistry was not one of Lamarck's strengths, merely to note that Lamarck had studied the subject sufficiently to wish to address certain issues. Two hundred years ago, physics and chemistry, as we know them today, were but young disciplines. The microscope was being refined. The organic cell had been seen but only as a 'blob' or globule. It was not known to contain anything within its walls other than protoplasm. As to the process of reproduction, that was still shrouded in mystery.

Both Buffon and Erasmus Darwin, in their respective writings about the Universe, had implied some depth of time. The Scottish geologist, James Hutton (1726-1797), with his interest in the formation of coal, had postulated great expanses of time, but his works were little known outside Edinburgh, in part due to his reclusive nature. His ideas had very little influence on the thinking of his fellow scientists and may well have been completely forgotten were it not for John Playfair who, in 1802, republished Hutton's work, together with a lengthy explanation, several times longer than Hutton's original papers! Lamarck would not have been aware of Hutton and the vast expanses of time which he believed necessary for evolution to have taken place would have been the product of his own thought, probably a gradual dawning over a number of years. While most thought the Earth was but a few thousand years old, Buffon and Erasmus Darwin (Charles Darwin's grandfather) probably thought in terms of tens or even hundreds of thousands of years. A careful reading of Lamarck's work makes it clear that Lamarck was thinking in terms of millions, although it is unlikely that even he conceived of billions. We will never know.

Lamarck started by telling his reader that his *Philosophie Zoologique* had come to be written as a result of his need to provide his students with a body of rules and principles which could be applied to the study of animals, paying particular attention to the differences apparent between families, orders and classes. Lamarck had noted that, as his studies took him 'downwards' from the most 'perfect' animals (as they were then known) to the least 'perfect' (those which fell within his Chair), there was a lessening of complexity: for in ascending the animal scale, starting from the most imperfect animals, organisation gradually increases in complexity in an extremely remarkable manner.

Lamarck observed that the simplest of organisms had no special organs whatever ... special functions arise ... and in the most perfect animals these are numerous and highly developed ... These reflections ... led me further to enquire as to what life really consisted of, and what are the conditions necessary for the production of this natural phenomenon and its power of dwelling in a body. Incredible as it may now seem, Lamarck was the first person to introduce the simple dichotomy inorganic/organic to replace the animal/vegetable/mineral categories then in use. That animal and plant life existed had been accepted as a fact. Lamarck was the first person to try to explain how and why life had come to manifest in previously inert matter. Of course, there had been many 'Creation' myths and the one in *Genesis* assumed, as did many others, than inert matter existed before 'life' took hold but quite why and how this had happened had not been the subject of scientific investigation. Attempts to produce 'life' in the laboratory have so far been unsuccessful, so the mystery of quite how and why it happened remains unsolved to this day.

Lamarck argued that our 'moral' (non material) being must have originated from the same source as our physical being, and our minds, intelligence, feelings and imagination must be subject to universal and eternal laws of the same exactitude as those which governed the appearance and development of our physical bodies.

Lamarck outlined the purpose and themes of his book on the second page:

The conditions necessary to the existence of life are all present in the lowest organisations, and they are here also reduced to their simplest expression. It became therefore of importance to know how this organisation, by some sort of change, had succeeded in giving rise to others less simple, and indeed to the gradually increasing complexity observed throughout the animal scale.

By means of the two following principles, to which observation had led me, I believed I perceived the solution to the problem at issue.

Lamarck's first principle was that of use/disuse. He argued that continued use of an organ developed and strengthened it, even enlarged it, while disuse caused the organ to deteriorate and even disappear. It was known that constant practice improved the voice of the singer, the manual dexterity of the musician, the strength of the blacksmith's muscles, while failure to practice, or work, caused deterioration. Many of these traits, and others, were seen to 'run in the family'. Quite how any characteristic was inherited was not then known but characteristics clearly were inherited and the concept of 'use/disuse' was commonly accepted. The first pillar of Lamarck's theory was built upon 'general knowledge'. He suggested nothing new. Use/disuse inheritance, as outlined by Lamarck, is sometimes referred to as the 'inheritance of acquired characteristics', but this term has also been used in another way, as will be discussed later. Lamarck's theory will be referred to as use/disuse inheritance, since that was the terminology he used.

Lamarck's first principle, therefore, was that change in environmental conditions would lead to change in the habit of living forms in that environment and this would lead to the development/deterioration of organs affected by that change.

Lamarck's second principle related to what he referred to as 'fluids' which moved within the body. The functioning of the nervous system was still poorly understood, although Lamarck was aware of experiments being carried out in this area, most involving newly discovered electrical currents. It must be remembered that gases, as well as liquids, are 'fluid' – they are capable of movement, unlike solids. The 'fluid' which was thought to operate the nervous system must not be thought of as a liquid. Nobody thought water, or blood, or some other 'liquid' flowed up and down the nerves. The 'fluid' was an as yet unidentified substance. Lamarck declined to speculate on its nature since nothing had yet been established. He believed that the fluids modify the cellular tissue in which they move, open passages in them, form various canals, and finally create different organs, according to the state of the organisation in which they are placed.

Lamarck based his whole theory upon the perceived interaction of the results of the movement of the 'fluids' within the body and the response of the body (via the action of the 'fluids') to changes in the environment.

Lamarck then addressed the 'property of feeling'. This property was not present in inanimate matter: I enquired therefore what the organic mechanism might be which could give rise to this wonderful phenomenon, and I believe I have discovered it. One can almost feel Lamarck's excitement as he penned those words!

Lamarck summarized the arguments he would put forward in his book thus: For the production of feeling the nervous system must be highly complex, though not so highly as for the phenomena of intelligence … the nervous system, when it is in the extremely imperfect condition characteristic of more or less primitive animals, is only adapted to the excitation of muscular movements, and that it cannot at this stage produce feeling. There were, Lamarck argued, no ganglia, no spinal cord the anterior extremity of which expands into a brain which contains the nucleus of sensation … When the nervous system reaches this stage, the animals possessing it have the faculty of feeling.

This paragraph is a good example of Lamarck's application of logic, rather than emotion, to the operation of nature. It is not easy at times to watch television documentaries in which some 'monster' of the undergrowth chomps its way through the body of a living victim. How cruel does Nature seem! The argument which would have been put forward by Lamarck, and which is put forward by some naturalists, is that these 'imperfect' creatures, with their less developed nervous systems, simply do not feel pain in the same way that do we. 'Humanizing' them, assuming that they fear death and endure untold suffering as they are eaten alive because that is what would happen to us, is not appropriate. Lamarck did not allow his emotions to determine what he thought creatures could, or could not, experience. He based his conclusions on the level of development of their nervous systems and of the various parts of their brains. Lamarck's observations led him to the conclusion that 'feeling' and 'irritability' were two different phenomena which did not have a common origin.

Lamarck had identified what today are known as the sensory and motor nervous systems, separate within the body but communicating with each other in the brain. Of course, 'imperfect' creatures were not devoid of all sensory perception. Many had eyes, some had a sense of smell vastly superior to that of humans, and all responded to touch. They had a capacity for 'irritability/perception.' Anyone who has received a numbing injection at the hands of a dentist will know that this gives relief from pain but does not prevent awareness of pressure as the dentist pushes and probes. This sense of touch/pressure exists at a 'lower', more primitive level than that of actual pain. Lamarck's observations related to the extremely imperfect condition characteristic of more or less primitive animals. When the nervous system progressed to the stage where a brain, however primitive, could be observed, that was the stage at which the faculty of 'feeling' first developed. Before this, animals were capable only of 'perception'.

Lamarck next addressed what he termed 'inner feeling': that feeling of existence which is possessed only by animals which enjoy the faculty of feeling … It is a feeling which can be aroused by physical and moral needs, and which becomes the source whence movements and actions derive their means of execution. No one that I know had paid any attention to it … the word emotion, which I did not create, is often enough pronounced in conversation to express the observed facts.

It is hard to appreciate today that this was a new concept, that animals had feelings which were capable of influencing their actions. According to *Genesis*, free will was the gift of God to Man alone. Animals were presumed to act by instinct, not to be capable of thought or emotion. On the premise that where there is no sense, there is no feeling, scientists had given themselves permission to experiment on live, and conscious, animals, going so far as to claim that the yelps and cries of the animals being dissected were but reflex actions. The practice of vivisection without anæsthetic persisted into the 20th century. It would almost certainly have been practiced at the Museum, particularly by medical students.

Lamarck concluded that plants and animals possessed only very primitive nervous systems, devoid of 'inner feeling' and the ability for conscious action, perceived only exterior excitation, which came to them via the "moving fluids contained in the environment [which] incessantly penetrate these organised bodies and maintain life in them…"

Lamarck combined this concept of 'inner feeling' with the two principles earlier outlined and recognized "the importance of this power in nature which preserves in new individuals all the changes in organisation acquired by their ancestors as a result of their life and environment." Originally, living forms, both plant and animal, merely responded to external influences, then Nature found a way to internalize the response so that groups of animals were capable of making group decisions, as did shoals of fish, flocks of birds and colonies of ants. Eventually Nature reached "the point of placing that same power at the disposal of the individual."

Lamarck went on to say:

The facts which I name are very numerous and definite, and the inferences which I have drawn from them appeared to me sound and necessary; I am convinced therefore that it will be found difficult to replace them by any others.

Notwithstanding this confidence, Lamarck expected his ideas to elicit opposition because that was the natural reaction to new ideas, one which he believed was beneficial, since it helped to ensure only sound ideas became established. Certain facts were irrefutable, all the rest were opinion and, no matter how secure each of us might feel in the correctness of our own opinion, we should be wrong to blame those who refuse to adopt our own. New ideas must first be a minority opinion and only with the passage of time become widely accepted, often referred to as 'common knowledge' or 'public opinion'.

There are then few positive truths on which mankind can firmly rely. They include the facts which he can observe, and not inferences that he draws from them; they include the existence of nature, which presents him with these facts, as also the laws which regulate the movements and changes of its parts. Beyond that, all is uncertain, although some conclusions, theories, opinions, etc., have much greater probability than others.

It is the above paragraph which most clearly distinguishes Lamarck's work from that of Charles Darwin. As already stated, by the time Darwin published *On the Origin of Species* in 1859, the concept of evolution was already widely accepted, but no one had offered a theory as to how or why change had taken place. As will be seen, Lamarck simply accepted that established characteristics were passed on to progeny but suggested no means by which these changes became established other than 'response to changed environment'. Darwin's theory of evolution by *Natural Selection* was speculation. He offered no 'scientific' evidence, or at least none that would pass the test of 'scientific' today, but that was not to be expected, since there was none. Even a cursory glance through *The Origins* will show how frequently Darwin prefaced his proposals by "I think ..." or "I believe ...". Of course, Darwin cited as many facts as he could to support his speculation, but it was all 'circumstantial'. In the 19th century, a prosecutor would assemble as many 'facts' as possible to present before the Court in the hope of carrying the case by weight of accumulated circumstantial evidence because, unless there was reliable eye witness testimony, that was all there was to rely on. Forensic evidence, as we understand it today, was simply not available. Darwin, simularly, set about accumulating circumstantial evidence to support his ideas from which he extrapolated to present his theory. Lamarck endeavoured to avoid speculation, choosing rather to apply deductive logic to such facts as he believed to be 'certain' to guide him to his conclusions. He emphasized his point by continuing:

We cannot rely on any argument, inference or theory ... There is nothing we can be positive about, except the existence of bodies which affect our senses, and of the real qualities which belong to them, and finally the physical and moral facts of which we are able to acquire a knowledge. The thoughts, arguments and explanations set forth in the present work should therefore be looked upon merely as opinions which I propose, with the intention of setting forth what appears to me to be true, and what may indeed actually be true … my purpose is to invite enlightened men who love the study of nature to follow them out, verify them, and draw from them on their side whatever conclusions they think justified.

Lamarck noted that he could have considerably extended his work by including all the interesting matter that it permits of but resisted the temptation so as not to overburden the reader. This was a problem for Darwin also, who famously spent nearly twenty years accumulating so much evidence that the great work which he planned to write was never completed. Both Lamarck's and Darwin's 'abbreviated' versions ran to about 400 pages.

Lamarck concluded his *Preface* by writing: I shall have attained my end if those who love natural science find in this work any views and principles that are useful to them … if the ideas which they succeed in giving rise to, whatever they may be, advance our knowledge or set us on the way to reach unknown truths.

PRELIMINARY DISCOURSE

THE purpose of this book is to draw attention to the work of Lamarck, what he actually wrote, what I believe he meant by what he wrote.

Lamarck's *Preliminary Discourse* was fairly short, less than seven pages, but in it he gave his reasons for writing his book and the philosophy which lay behind it. Over the course of his life, largely due to the nature of his work, Lamarck had become introspective and developed a strong philosophical outlook which underpinned all his work and thinking. He felt it necessary to rewrite both his *Système des animaux sans vertèbres* and his *Recherches sur les corps vivants*, offering his current work, not just as a body of scientific knowledge, but as a scientific philosophy.

Understanding this chapter is crucial to understanding Lamarck's philosophy. Lamarck rejected the separation of science and religion/ philosophy, insisting that a philosophy of science is not only possible, but inevitable and unavoidable. To observe nature ... to endeavor to grasp the order which she everywhere introduces ... her laws ... these are in my opinion the methods of acquiring the only positive knowledge that is open to us. I believe that what Lamarck was saying was that the study of Nature was a more sure guide to knowledge, including knowledge of our place in the Universe, than was the study of the Bible. Lamarck did not mention the Bible, said nothing against it, but not mentioning it in itself was significant, because some mention of God, of his truth as taught in the Bible, was to be expected. Lamarck's complete avoidance of any mention of God or the Bible probably told more about Lamarck's underlying philosophy than any mention would have done, since it would have come as a surprise, possibly even a shock, to his reader.

The next sentence is bitter-sweet, bearing in mind the difficulties Lamarck endured in his final years: It is at the same time the means to the most delightful pleasures, and eminently suitable to indemnify us for the inevitable pains of life.

And in the observations of nature, what can be more interesting than the study of animals? This comment came, remember, from one who had devoted a large part of his life to the study of plants.

Lamarck outlined four questions to which he had sought answers, although he did not number them, as I am doing here. They were:

> 1) The question of the affinities of animal organisation with that of man;
> 2) The question of the power possessed by animals to modify their organs, functions and habits according to their modes of life, climates and places of habitation;
> 3) The question of the different systems of organisation which guide us in determining relationships within the scheme of nature;
> 4) The question of general classification according to greater/lesser complexity of organisation.

In *On the Origin of Species*, Darwin studiously avoided the topic of human evolution, for which he was criticized, the similarity of form between humans and apes being so obvious. Lamarck had no such hesitation, the affinity of humans with animals being the first question he listed for consideration. From there he went on to the question of change (evolution) itself, then that of the relationship of the different orders of animals which varied so greatly and, lastly, to degree of complexity, which term he preferred to use rather than perfect/imperfect, since each creature was perfect in its own way.

These questions, claimed Lamarck, were of very great interest to anyone who loves nature and seeks the truth in all things. This comment, together with similar comments made in the Preface, show how much Lamarck loved and respected Nature. In the past, Naturalists had chosen to study primarily mammals, birds, reptiles and fish. These, which Lamarck had now termed 'vertebrates', were mostly larger than the invertebrates and had their parts and functions better developed. Only now were Naturalists setting aside their former prejudice, which had led to the invertebrates being despised by the vulgar (common person, non aristocrat, non professional). During the few years that the invertebrates had been studied, they had come to be acknowledged as extremely interesting since they shed new light on a number of problems: The most important discoveries of the laws, methods and progress of nature have nearly always sprung from the examination of the smallest objects which she contains … My researches … soon inspired me with the highest interest in the subject.

Not only were the invertebrates more numerous than the vertebrates, they were far more varied, many in quite remarkable ways, but the successive development of different organs was more clearly to be seen. The best approach, Lamarck believed, was first to take a broad overview of the whole spectrum, then to divide it into parts, then into sub-parts, finally arriving at the smallest living part. Microscopes were becoming increasingly powerful, although at this time they were still not powerful enough to distinguish parts within the cell, the cell being the smallest part to which Lamarck was able to direct his attention. The precise method of reproduction was still unknown.

Lamarck was of the opinion that established methods had concentrated too much on the form of the entity being studied and not sufficiently upon its nature, its relationship to other forms and the reason for modifications and variations. Since our physical and our moral attributes stem from the same source, they must interact. As physical forms have evolved and become more complex, the separation of the two aspects of our being has become more apparent, to such an extent that some considered that they had nothing in common. Lamarck asserted an error had been made in studying first human beings, in whom the two strands were most developed (most complex). He believed that the close association, the interaction of the one strand upon the other, would have been more clear if scientists had started their study at the other end of the spectrum, with that of the most simple living entities and their development into what appeared first to be conscious beings and, later, beings with some degree of free will.

Lamarck concluded his *Preliminary Discourse* by outlining the structure of his book, which fell into three parts. The first part was concerned with what he called *artificial devices,* such as the division of living things (both plant and animal) into species. He would then, starting with the most complex animals, mammals, show the degradation running through the animal kingdom apparent upon studying the lesser animals and show the influence of environment and habit on the organs of animals, as being the factors which favour or arrest their development. He would conclude the first part of his work by suggesting their most suitable arrangement and classification.

In the second part, Lamarck would discuss what constitutes the essence of animal life; and ... the conditions necessary for the existence of this wonderful natural phenomenon. He would then endeavor to ascertain the exciting cause of organic movements ... the properties of cellular tissues ... the sole conditions under which spontaneous generation can occur.

The third and final part of his work would discuss the physical causes of feeling, of the power to act, and the acts of the intelligence found in certain animals. This would include:

1) origin and formation of the nervous system;

2) the nervous fluid;

3) physical sensibility ... sensation;

4) the reproductive power of animals;

5) the origin of will and the faculty of willing;

6) ideas and their different kinds;

7) understanding, attention, thought, imagination, memory, etc.

This brief overview of what was to come is sufficient to show that this was to be a major scholarly work of a breadth and depth never before attempted, even by Buffon.

PART 1
The Natural History
of Animals

Chapter I – Artificial devices

L AMARCK commenced by pointing out the necessity of organizing any large undertaking into manageable portions. These portions would need to be identified in some way. However logical and practical may be the process involved, the resultant classifications would nevertheless be artificial devices. Lamarck stressed the importance of always keeping this fact in mind, of the natural scientist never allowing himself to think or believe that the classifications used to identify plants and animals had any reality in nature. They did not constitute the laws and acts of nature herself ... among her productions nature has not really formed either classes, orders, families, genera or constant species, but only individuals who succeed one another and resemble those from which they sprung.

It is interesting that Darwin commenced his argument in *On the Origin of Species* the same way. The difference between the two men was that, while Lamarck focused on the higher classifications, classes, orders, families, genera and species to illustrate his arguments, Darwin built his theory on individual variation. Some, said Lamarck, studied nature for its economic interest; their researches were aimed at discovering what there was in nature which was of practical and/or economic value to them and/or their community. Others studied nature out of philosophic interest through which we desire to know nature for her own sake. Such people were the only true scientists since they brought an impartial attitude to their studies.

Lamarck praised the work of Linnæus but felt that his good work was being undermined by a tendency on the part of natural scientists to laxness in the application of the rules of classification. Many were introducing sub-classes, sub-orders, sub-families and sub-genera, which practice was causing confusion. The ability to communicate one's work, one's thoughts and ideas, to fellow workers required communality of language. Briefly, Lamarck defined the classifications thus:

Classes

Highest general division. Although some classes, such as mammals and birds, appeared to be really marked out by nature herself ... this is none the less a pure illusion, and a consequence of the limitation of our knowledge of existing or past animals. Lamarck was preparing his reader for the concept of descent from a single source which, logic demanded, would have necessitated a blurring of apparently clear divisions at some time in the past. Animals were being discovered in Australia which appeared to blur boundaries. How many were there yet to be discovered, on land, beneath the sea, living or fossil? Lamarck argued that it was inappropriate to set up class within class; that, however, is just what has been done ... We shall soon reach not only sub-classes but sub-orders, sub-families, sub-genera and sub-species. Now this is a thoughtless misuse of artifice, for it destroys the hierarchy and simplicity of the divisions, which had been set up by Linnæus and generally adopted … A stop must be put to the abuse of nomenclature.

Orders

Orders are the main divisions of the first rank into which a class is broken up … the class of molluscs, for example, are easily divided into two groups, one having a head, eyes, etc., and reproducing by copulation, while the other has no head, eyes, etc., and carry out no copulation to reproduce themselves. Cephalic and acephalic molluscs should be regarded as the two orders of that class. The orders into which a class is divided should be determined by the presence of important characters extending throughout the objects comprised in each order.

Families

The name family is given to recognized parts of the order of nature … smaller than classes but larger than genera … the boundaries of these families are always artificial … Some divide one family into several new ones and others combine several families into one. **Today these two groups of theorists would be known as 'splitters' and 'lumpers'.**

The number of new plants and animals then being discovered led Lamarck to claim that there are many races of animals and plants that are still unkown to us … The gaps thence arising … will leave us for a long time still, and perhaps for ever, [without] the means for setting up [securely] the majority of the divisions … Alterations in the boundaries, extent and determinations of families will always cause a change to their nomenclature.

Genera

Lamarck introduced *genera*, not as a subdivision of *family*, but as combinations of races or so-called species that have been united on account of their affinities, and constitute a number of small series marked out by characters arbitrarily selected for the purpose. When a genus is well made, all the races or species comprised in it resemble one another in their most essential and numerous characters. They differ only among themselves in characters less important, but sufficient to distinguish them.

Darwin built his whole theory of evolution by natural selection upon the lowest division, claiming that *species*, *race* and *variety* were interchangeable terms, having no reality in nature. Some varieties were 'well-marked' and usually referred to as 'species'. Others, less 'well-marked' were usually termed 'varieties' but, which plants or animals were true species and which were varieties, was subject to constant debate.

The constant change which troubled Darwin, had troubled Lamarck half a century before. Lamarck appealed for naturalists to agree to make the study of nature simpler for them all. He believed that the higher classifications, from *kingdoms* to *genera*, did exist in nature, for all practical purposes, but that *species* and *varieties* were *artificial devices*, which, while having an *indisputable utility*, were devices whose contrived nature should always be borne in mind.

Chapter II – Affinities

I T WAS agreed among naturalists that the closer and more extensive the resemblance between two living bodies, the greater their affinity, particularly when the resemblance referred to essential parts. How great has been the progress of natural science since serious attention began to be given to affinities, and especially since their true underlying principles have been determined? … In the animal kingdom the invertebrate animals comprising the larger part of all known animals were classified into the most heterogeneous groups, some under the name of insects, some under the name of worms.

The principle of natural affinities removes all arbitrariness from our attempts at a methodological classification of organised bodies … Naturalists are forced to agree as to the rank which they assign … they are obliged to follow the actual order observed by nature in giving birth to her productions…

If the affinities are so great that not only the essential parts, but also the external parts present no determinable differences, then the objects in question are only individuals of the same species. If … the external parts exhibit appreciable differences, though less than the essential resemblances, then the objects in question are different species of the same genera.

The proportions and relations of the parts of individuals composing a species or a race always remains the same, and so appear to be preserved for ever … changes only take place with an extreme slowness, which makes them always imperceptible … Hence when he [the observer] comes across any species which have undergone these changes, he imagines that the differences which he perceives have always existed.

The most important parts for exhibiting the chief affinities are, among animals, the parts essential to the maintenance of life, and among plants to reproduction. In animals, therefore, it is always the internal organisation which will guide us in deciding the chief affinities. Among animals … three kinds of special organs have rightly been chosen … as the most suitable for disclosing the most important affinities:

(1) The organs of feeling. The nerves which meet at a centre, either single as in animals with a brain, or multiple, as in those with a ganglionic longitudinal cord;

(2) The organs of respiration. The lungs, the gills and tracheæ;

(3) The organs of circulation. The arteries and veins, which usually have a centre of action in the heart.

The first two of these organs are more widely used by nature, and therefore more important than the third … the organs of circulation disappear in the series after the crustaceans, while the two former extend to animals of the two classes which follow the crustaceans … The organ of feelings has … more importance from the point of view of affinities … without that organ muscular activity could not take place.

Lamarck then made a few comments about affinities of plant parts. The next paragraph sums up the fifteen years (or more) of work that had led Lamarck to form his philosophy of zoology, which he was now presenting:

It was, in fact, due to the perceptions of the importance of affinities that the attempts of the last few years were originated to determine what is called the natural method; a method which is only a tracing by man of nature's procedure in bringing her productions into existence. **Lamarck went on to say that** it was through affinities that I recognized that infusorian animals could no longer be put in the same class as polyps; that radiarians also should not be confused with polyps; and that soft creatures, such as medusæ and neighboring genera, which Linnæus and even Bruguière placed among the molluscs, were essentially allied to the echinoderms, and should form a special class with them.

It was again the study of affinities which convinced me that worms were a separate group, comprising animals very different from the radiarians and still more from polyps; that arachnids [spiders] could no longer be classed with insects, and that cirrhipedes were neither annelids nor molluscs.

The 'worms' of Linnæus were the 'critters that crawled' and the insects were those that flew. The vastness of their numbers and the magnitude of their variety had made classification seem impossible. Lamarck realized that by carefully and accurately determining the provisions for each of the highest classifications, starting with classes, then orders, then families, it would be possible to bring this apparent chaos into order, since chaos did not occur in nature, only in human lack of understanding. Specialists in the study of particular families would allocate genera and, after that, species, although this final division would always be problematic, since it was artificial.

It must be remembered that, at the time Lamarck was writing, there was no understanding of genetics. It was known that 'each reproduced after its own kind', but how this was accomplished was not known. The inability of two different species to interbreed due to genetic incompatibility was then unknown.

Lamarck noted a few other changes he had made and commented that when a similar study of affinities among plants had been made, then the classification of living bodies would no longer be left to arbitrary judgment.

Chapter III – Species

HAVING completed his preliminary observations, Lamarck commenced Chapter III by addressing the central issue at the heart of evolutionary theory, the same issue which is still being debated today by Darwinists, Creationists and Intelligent Design theorists. He asked if it is true that all species are of absolute constancy, as old as nature, and have all existed from the beginning just as we see them today; or if, as a result of changes in their environment, albeit extremely slow, they have not in course of time changed their character and shape.

Lamarck defined species as any collection of like individuals which were produced by others similar to themselves. Lamarck claimed that this definition was only exact in the short term, that over time species did change in response to changes in their environment. Species have really only a constancy relative to the duration of the conditions in which are placed the individuals composing it. Anticipating the identical argument put forward by Darwin, he claimed that naturalists made arbitrary decisions regarding whether a population of plants or animals be considered a species or a variety, that there was often disagreement between naturalists on this issue and that, therefore, changes and revisions were constantly being made, to the confusion of naturalists and the detriment of their science.

Like other naturalists of his time, Lamarck was profoundly impressed by the way in which the discovery of new animals, especially in Australia, appeared to fill the gaps between species: the richer our collections grow, the more proofs do we find that everything is more or less merged into everything else. The monotremes, being egg-laying mammals, seemed to bridge the gap between birds and mammals, even if they had no feathers; the external pouch of the kangaroo seemed likewise to offer a basic form of uterus, albeit outside the mother's body, but allowing the infant to suckle. That a great southern land existed had been known for hundreds of years, although it was not until the 17th century that European nations became interested in mapping it. The English, the French and the Dutch in particular explored this part of the southern ocean, the Spanish having staked their claim to South America, at that time the other great unknown land. Captain Cook mapped much of Australia's coastline in 1770 and England annexed parts of Australia in 1776, although full annexation did not occur until 1856. At the time Lamarck was writing, France, in the person of Captain Baudin, was exploring Australia with a view to the acquisition of territory. The novelty of Australia's fauna was a matter of great scientific interest, reports of which were almost unbelievable until specimens finally arrived in Europe.

All ships of exploration carried naturalists, most of whom were also proficient artists, and detailed accounts of new flora and fauna were sent 'home', avidly awaited and devoured. Whenever possible, dried samples were sent. By the time Darwin returned from his five year travels the situation had changed. The museums were so full of specimens that he had difficulty finding any to accept his offerings and he was forced to sort and categorize much of his collection himself – with the help of some friends.

Lamarck and Cuvier would have been the first to examine both the dried specimens and the written/drawn accounts reaching the Museum. Imagine the excitement of these two men as they became aware, not merely of new varieties or species of a genera, but of previously unknown families and classes, which it fell to them to describe and classify! In addition to new living animals, the Museum was also identifying and, where possible classifying, fossil remains which were being found in increasing abundance, due to the building of better roads, canals and railways. What an exciting time to hold one of the two Chairs in Zoology at this prestigious organization!

Lamarck explained: I do not mean that existing animals form a very simple series, regularly graded throughout; but I do mean that they form branching series, irregularly graded and free from discontinuity, or at least once free from it. For it is alleged that there is now occasional discontinuity, owing to some species having been lost … How great the difficulty now is of studying and satisfactorily deciding on species among that multitude of every kind of polyps, radiarians, worms, and especially insects, such as butterflies, Phalæna, Noctua, Tinea, flies, Ichneumon, Curculio, Cerambix, chafers, rosechafers, etc! These genera alone possess so many species which merge indefinably into one another. What a swarm of mollusc shells are furnished by every country and every sea, eluding our means of distinction and draining our resources.

This was the first time that Lamarck clearly drew upon his speciality, the invertebrates, to justify his thinking. Was it not the same with fish, reptiles, birds, mammals and plants? Lamarck held that this situation had come about by gradual change in response to a changing environment and drew upon the changes which were known to occur under domestication to support his claim. When conditions stayed the same, living bodies stayed the same. When conditions changed, and remained changed for a lengthy period of time, then after a long succession of generations these individuals, originally belonging to one species, become at length transformed into a new species distinct from the first.

Suppose, for example, that the seeds of a grass or any other plant that grows normally in a damp meadow, are somehow conveyed first to the slope of a neighbouring hill where the ground although higher is still rich enough to allow the plant to maintain its existence. Suppose that then, after living there and reproducing itself many times it reaches little by little the dry and almost barren ground of a mountain side. If the plant succeeds in living there and perpetuating itself for a number of generations, it will have become so altered that botanists who come across it will erect it into a separate species.

This last paragraph is notable as being one of only two in the entire book in which Lamarck invoked supposition, a technique utilized by Darwin to such an extent that this paragraph could seamlessly be inserted into *The Origins* and it is unlikely that anyone other than an expert would notice the insertion or the change in authorship. This was Darwin's natural selection.

Plants were able to change more quickly than animals. We now understand that some plants are able to reproduce asexually, relieving them of the necessity of mating with a partner whose DNA is virtually identical with their own. A spontaneous alteration in their genetic material has more chance of being viable than in a sexually reproducing plant or animal.

Lamarck then disputed the assumption that individuals of one species could not unite in reproductive acts with individuals of another species ... this assumption is unwarranted; for the hybrids so common among plants, and the copulations so often noticed between animals of very different species, disclose the fact that the boundaries between these alleged constant species are not so impossible as had been imagined.

Like Lamarck, Darwin also based his theory of evolution on the claim that 'species' was an artificial distinction, invented by naturalists for their own convenience, which had no validity in nature. Darwin, also, cited hybrids in support of his position. The best known hybrid is that of the mule, the offspring of a horse and a donkey. Mules are strong and healthy, appear perfectly formed although somewhat different from either parent, but have one peculiarity: they are infertile. Why this should be the case was a great puzzle to naturalists before the era of modern genetics. We now know that, while the horse has 62 chromosomes and the donkey 64, their DNA is sufficiently close for them to be able to produce a mule, which has 63 chromosomes, making it infertile.

All living forms, be they plant or animal, which reproduce sexually, must have an even number of chromosomes, since the number is reduced to half during the reproductive process, being restored to the full complement on completion of fertilization, half the genetic material being provided by each parent. Darwin suggested that mules did not breed because the conditions in which they were kept were not favourable. It was known that healthy animals, such as lions, failed to breed in captivity because the conditions of their captivity were not favourable, so this was not an unreasonable suggestion. Lamarck and Darwin both theorized that hybrids thus formed could/would over time give rise to new varieties and then to new species.

Lamarck stated that only the simplest organisms were created directly by nature – were *spontaneously generated*. Everything else had developed over time. The property of growth is inherent in every part of the organised body, from the earliest manifestations of life.

Lamarck recognized that the complexity of living forms would cause many to postulate the involvement of a Supreme Author of all things:

No doubt he would be a bold man, or rather a

complete lunatic, who should propose to set limits to the power of the first Author of all things; but for this very reason no one can venture to deny that this infinite power may have willed what nature herself shows us it has willed … should I not recognize in this power of nature, that is to say in the order of existing things, the execution of the will of her Sublime Author, who was able to will that she should have this power?

Shall I admire the greatness of the power of this first cause of everything any the less if it has pleased him that things should be so.

There are many today who hold a similar opinion, who believe that some form of Divine Being or Creative Force created the Universe and, having set its natural laws in motion, has allowed it to evolve and function ever since, without further interference. At the time that he wrote *On the Origin of Species*, Darwin held a similar view. Others fail to understand why the Divine Being's creative energy should all have been expended in the one act of creation and see the Creator as being onwardly involved in His creation. They argue that evolution may explain change but fails to explain the origin of anything. Another group are the atheists, the humanists, who deny the existence of any form of God or Creative Being, denying that there is any form of Intelligence superior to that of the human being. These people accept evolutionary change over great expanses of time but believe the Universe, and its natural laws, to be self created. On the opposite side of the spectrum are the Creationists, or the Special Creationists as they are now being called. These people accept God but deny evolution. They are mostly Fundamental/Orthodox Jews/Christians/Muslims of the Abrahamic tradition, who accept the account of Creation given in the book of Genesis in the Bible. Theoretically, there should be one more group, those who reject both God and evolution, who believe that the Universe is self created and that the world exists today as it did when it first came into being. However, I have never come across such a person and it is likely, therefore, that you, the reader, find yourself relating to one of the first three groups mentioned.

Lamarck closed his chapter on species by suggesting that, other than at the hand of man, no species had ever gone extinct, but had simply changed its form. Lamarck visualized change taking place very slowly in response to slow changes in the environment. Lamarck felt that the effects of earthquakes and volcanos were very local and, while they may cause the death of some individuals in their locality, they were unlikely to cause the death of an entire species. Similarly, climate and environmental changes would not occur sufficiently rapidly to kill an entire species. The species would adapt to the changed conditions, they would evolve over millions of years. Nothing on the surface of the earth remains permanently in the same state. Ground is constantly being elevated or depressed, rain and river are weathering the rocks, and living entities are responding to these changes which, of course, include changes in each other as predator and prey! Rather than wondering at finding fossils different from species living today, Lamarck felt rather that we should wonder that any are found similar to forms which existed in the distant past.

Chapter IV – Animals

LAMARCK was meticulous. Before presenting his theory of the evolution of animals, he first considered it necessary to define the term 'animal'. However varied their size, shape and organization, animals had one thing in common, the ability to move in however small a way, without external stimulus. Unlike plants, which were able to absorb the nutrients necessary for their survival from their immediate surroundings, animals must actively acquire nutrients. Most needed actively to find a mate. The few not capable of movement for this purpose had to be provided by their environment with facilities for this to occur, although it must be said that this is also true of sexually reproducing plants, so this is not as clear a distinction between plants and animals as the two previous points. Lamarck defined the properties of animals as follows:

> (1) Some only move themselves or their parts when their irritability has been stimulated; but they experience no feeling: these are the most imperfect animals...
>
> (2) Others … are capable of experiencing sensations, and possess a very vague inner feeling of their existence; but they only act by the internal impulse of an inclination...
>
> (3) Others … not only are they capable of receiving sensations, and possess an inner feeling of their existence, but they have besides the faculty of forming ideas... and of acting by free will, subject however to inclinations which

lead them exclusively towards certain special objects...

(4) Others ... are able to form clear and precise ideas of the objects which affect their senses and attract their attention; to compare and combine their ideas up to a point; to form judgments and complex ideas; in short to think, and to have a will...

The most imperfect animals had no energy for movement, only enough for vital movements, such as the opening and closing by the oyster of its shell 'mouth' for the absorption of food. As the amount of vital energy present increased, so, too, did the complexity of the animal's organization. Muscular activity was essential for movement, but this necessitated both a circulatory system and a nervous system along which the necessary 'fluids' might pass. The reader will remember that both liquids and gases are fluid and these were all which it was then believed were available to the body for its function. That the body also hosted electrical impulses was not then established, although the *detrimental* effect of electricity upon the body was. Breath was essential for life and it was assumed that some form of gas, such as oxygen and possibly others, passed into the blood stream and throughout the body, being utilized by the nervous system, as well as the circulatory.

Lamarck claimed that the most important differences between the various classes, orders and families of animals lay not in their varied shape, size and faculties, but in their internal organization. A tiny insect, able to move each of its six legs, able to fly, to choose where it will land for the purpose of feeding, of seeking for itself a mate, is far more complex, far more 'perfect', than an oyster many times its size, which can but sit all day, attached to a rock, and merely open and close its mouth passively to receive food.

All the organs, even the most important, arise one after the other in the animal scale, and afterwards become successively more perfect through the modifications impressed on them … The examination of the internal organisation of animals … is then the subject of study most deserving of our attention … animals … are rendered extremely remarkable by their faculty of locomotion, a great many of them still more so by their faculty of feeling …

The faculty of feeling is still very obscure and limited in the animals among which it begins to exist; but it then develops gradually, and when it has reached its highest development, it ultimately gives rise in the animals to the faculties which constitute intelligence.

Indeed the most perfect among animals have single and even complex ideas; they have passions and memory and they dream … they are up to a certain point capable of learning … Nature thus succeeds in endowing a living body with the faculty of locomotion, without the impulse of an external force; of perceiving objects external to it; of forming ideas by comparison of impressions received … of comparing or combining these ideas, and of forming judgments which are merely ideas of another order; in short of thinking.

… an enormous time … must doubtless have been required to enable nature to bring the organisation of animals to that degree of complexity and development in which we see it at its perfection.

Lamarck referred to the three kingdoms, animal, vegetable and mineral, into which nature had traditionally been divided. He suggested that nature is, in fact, divided into two main branches:

1) Organised living bodies;
2) Crude bodies without life.

These two branches are organic and inorganic.

Living beings, such as plants and animals, constitute the first of these two branches … they possess … the faculties of alimentation, development, reproduction and they are subject to death.

What is not known as well … is that living bodies form for themselves their own substances, as a result of the activity and functions of their organs … What is still less known is that the exuviæ of these living bodies give rise to all the composite matters, crude or inorganic, that are to be found in nature, matters of which the various kinds increase in course of time and according to the conditions, by reason of the disintegration which they imperceptibly undergo. For this disintegration simplifies them more and more, and after a long period leads to the complete separation of their constituent principles.

It was this paragraph which Elliot interpreted as indicating that Lamarck believed that all inorganic matter had its origin in organic matter and it is agreed that, taken completely literally, Lamarck's statement that the exuviæ of living bodies gives rise to all composite matters, crude or inorganic, could be so construed. However, I refuse to believe that Lamarck seriously thought that living bodies, both plant and animal, evolved in water and on land when neither water nor land existed! How would animals have developed their means of locomotion if there was no earth for them to move upon? How would plants have grown out of the ground if the ground did not exist?

Even without the further explanation which occurs later in his book, I believe that a sensible reading of this passage would consider first Lamarck's reference to 'composite matters'. Lamarck had realized that living bodies, be they plant or animal, were chemical factories, which converted nutrients into new substances not found in the earth. These composite substances eventually returned to the earth, one way or another. Most plants and animals were eventually consumed, either killed and consumed in their entirety, partially consumed like many plants, or decomposed after death. Only bones and shells had a chance of surviving into fossil material. Much of the earth beneath our feet was of organic origin, such as chalk, clay and soil. Chalk cliffs might rise for hundreds of feet out of the earth, might extend for hundreds of feet beneath it, might have taken countless years to form beneath the sea from the remains of tiny sea creatures and countless more to be raised up above it again, but they owed their origin to creatures which lived millions of years ago. Think of the marble which graced European architecture and art from the time of the Ancient Greeks and Romans through to the Renaissance and beyond. Formed at such great depths beneath the sea that the limestone had been compressed and turned into solid stone, yet that stone had been formed from the remains of living creatures.

There must have been some inorganic rock which constituted the original Earth. I do not believe that Lamarck was suggesting that volcanic material was of organic origin. However, what he was drawing attention to was the change which had taken place on the surface of the Earth, not just as the result of earthquakes and volcanoes, but as the result of life itself. All the soil beneath our feet, which sustains the plants which feed the animals, on which we cultivate our crops and pasture our herds, is the product of life. Lamarck was aware of an eternal cycle of life and death, where that which was alive returned to an inanimate form only to resurrect once more and live again. The ground beneath his feet was part of the cycle of life. Had he lived a couple of hundred years later than he did, he would have found affinity with the concept of Gaia, but he was a man ahead of his time.

Much of the composite matter formed within living bodies returned to the earth as excrement, not only excrement of large animals, but even of small insects and other creatures who live, mostly unnoticed by us, devouring dead leaves, fallen branches, dead bodies and, indeed, excrement itself. Gradually organic matter is broken down. Even bone which is not eaten is leached of its minerals by weathering. Everything gradually returns to the earth, from where it is reabsorbed by plants and eaten by animals in an eternal cycle.

Lamarck concluded by saying: These are the various crude and lifeless matters, both solid and liquid, which compose the second branch of the productions of nature, and most of which are known under the name of minerals. I concede that Elliot could argue from this that Lamarck held that living bodies existed first and that minerals and other inorganic matter followed. However, I believe that when Lamarck referred to "the productions of nature", he was referring to that which had appeared on the surface of the Earth since life was generated thereon. He was wrong about the origin of minerals, although right in believing that they were to be found in living bodies, albeit in a different form. Our bodies are rich in iron, phosphorus and, of course, calcium, as well as other minerals. Scientists define 'organic' as a substance containing carbon, which is only found in material which is living or has once lived. In that one case, Lamarck appears to have been correct in claiming that a 'mineral' was formed by living bodies.

I believe that in his mind, Lamarck was going back into the mists of time, to when Earth had once been lifeless, and was trying to visualize how life came to exist and the changes which that life had wrought. I do not believe that Lamarck was stupid enough to envisage that living creatures existed suspended in nothingness until their dead bodies created solid ground. Nevertheless, Lamarck could have expressed himself better.

Lamarck concluded this chapter by a consideration of plants, which he claimed had developed separately from animals from the very beginning of life. Plants had no capacity for internal sensitivity such as would produce even the most primitive form of action initiated from within the plant itself. All action observed in plants was the result of external stimulus, and that included the remarkable 'folding' of the leaves of the 'sensitive plant' (Mimosa pudica). Movement occured in animals as the result of the contractibility of muscular fibres. This ability was completely absent in all plants. Having made a careful study of mimosa, Lamarck had concluded that, upon being touched, the 'branches and petioles' relaxed, indeed collapsed. Repeated touching elicited no further reaction. He noted that all plants with an apparent ability to move in some degree, although none other as much as mimosa, grew in hot climates. He postulated that some form of gas was given off during their respiration which accumulated in the joints between their leaves and stems. Upon being touched, this gas was suddenly released – rather, I assume, like the gas from a burst balloon. It took a long time for further gas to be produced and exhaled and for the petioles to return to their former position. Lamarck saw no similarity between this reaction and the voluntary movement of animals. All movement of plants was purely reactive. The opening and closing of the petals of flowers was stimulated by the light and warmth of the sun. The approach of darkness caused the dissipation of the accumulated gases and allowed plants to close their petals in 'sleep'.

As the result of his deliberations, Lamarck offered the following definitions of animals and plants:

Animals

Animals are organized living bodies, which have irritable parts at all times of their lives; which nearly all digest the food on which they live; and which move, some by acts of will, either free or dependent, and others by stimulated irritability.

Plants

Plants are organized living bodies whose parts are never irritable, which do not digest or move either by will or true irritability. Lamarck observed that a single class of invertebrate animals, such as insects for instance, equals the entire vegetable kingdom in the number and diversity of its contained objects. The class of polyps is apparently much more numerous still, but we shall never be able to flatter ourselves that we know all the animals which make it up.

Lamarck concluded by commenting on the fact that rapid rates of reproduction, especially among the smaller species, would result in the earth rapidly becoming overpopulated were it not for the fact that everything was constantly eating everything else. Species rarely ate their own kind, but made war on other species. Humanity alone, not being unduly threatened by any other species, was left with the responsibility of containing its numbers by making war on its own kind. The words of a soldier!

Chapter V – Classification of Animals

LAMARCK commenced Chapter V by reiterating that the artifice of classification was essential to assist us in our understanding of the general arrangements which occur in nature but it may be subject to re-evaluation, unlike affinities, which are actual works of nature and, therefore, unchanging. Our attempts at classification must be based, as far as possible, on these natural affinities.

Every class should comprise animals distinguished by a special system of organisation. The strict execution of this principle is quite easy and attended with only minor inconvenience… only a small number of animals admit … doubt as to their true class …

Each distinct group has its special system of essential organs; and it is these special systems which undergo a degradation as we pass from the most complex to the simplest. But each organ taken by itself does not proceed so regularly in its degradations … the organs that have little importance or are not essential to life are not always at the same stage of perfection or degradation …

These irregularities in the perfection and degradation of inessential organs are found in those organs which are the most exposed to the influence of the environment… species often constitute lateral ramifications around the groups to which they belong.

While on the one hand Lamarck was very aware of the similarities (affinities) which occurred within families, orders and classes, yet on the other hand he was also very aware of the differences between them. There must be similarities and points of cross-over, since Nature has fashioned all life forms from the same basic material and form. In cases of necessity nature passes from one system to another not merely in two different families but in one individual ...

Those systems of organisation in which respiration is carried on by true lungs are nearer to the lungs in allied classes and families, as is seen among fishes and reptiles, but she does so even during the existence of one individual; which possesses in turn first one and then the other system ... the frog in its imperfect condition of tadpole, breathes by gills; while in the more perfect condition of frog it breathes by lungs. But nowhere does nature pass from the system of tracheæ to the pulmonary system.

Insects, for example, breathe by tracheæ, by tiny 'tracks' or passages, which penetrate the body, allowing air to diffuse into the tissues. At no stage of their life are insects possessed of gills or lungs, even though many undergo quite striking metamorphosis.

In each kingdom of living bodies the groups are arranged in a single graduated series, in conformity with the increasing complexity of organisation and the affinities of the object. This series ... should contain the simplest and least organized of living bodies at its anterior extremity and end with those whose organisations and faculties are most perfect.

The first rule to be established, therefore, is that one extremity of an order must comprise the most complex (perfect) beings and the other the least complex (least perfect). Lamarck made a few comments on the arrangement of plants before returning to animals and summarizing:

The mammals will of necessity occupy one extremity of the order, while the infusorians will be placed at the other … We were obliged to recognize this order in each kingdom of living bodies … We shall now see that in the animal kingdom it is established in its outline in a way that leaves no scope for arbitrary opinions …

Aristotle indeed divided animals primarily into two main divisions or, as he called it, two classes, viz:

1) Animals that have blood:

Vivaporous quadrupeds

Oviparous quadrupeds

Fishes

Birds

2) Animals that have no blood:

Molluscs

Crustaceans

Testaceans

Insects

Lamarck did not immediately draw attention to the fact that the first group was the same as his vertebrates. He commented that:

This primary division of animals into two main groups was fairly good, but the character taken by Aristotle for discrimination was bad … He imagined that all animals which he placed in his second class only possessed white or whitish fluids; and he thereupon regarded them as having no blood … That erroneous direction has generally been followed ever since in the arrangement of animals; and this has clearly retarded our knowledge of nature's procedure …

Modern naturalists have endeavoured to improve upon Aristotle's division by giving to the animals of the first class the name of red blooded animals, and to those of his second class that of white-blooded animals. It is now well known … there are some invertebrate animals (many annelids) which have red blood. Here, at last, Lamarck makes the association between Aristotle's two classes and his own of vertebrate and invertebrate, which was sound and has stood the test of time. Lamarck chose to restrict the term 'blood' to fluid which circulated in arteries and veins, pointing out that other fluids were so varied that one might as well attribute blood to a plant as a radiarians or polyps using any other definition.

Lamarck then made the definitive statement that since his lectures given in the spring of 1794 he had divided animals into two distinct groups: those that have vertebræ and those that do not. The whole plan of organization of vertebrate animals was different from that of invertebrates. Lamarck then listed Linnæus' six classes of animals:

Classes	First Stage
I Mammals	Heart with two ventricles;
II Birds	blood red and warm
	Second Stage
III Amphibians	Heart with one ventricle;
(reptiles)	blood red and cold
IV Fishes	

                          Third Stage
        V Insects   A cold serum
        VI Worms   (in place of blood)

The first four of Linnæus's divisions were now definitely established. This was not the case with the last two divisions – the invertebrates which Lamarck had made the area of his special expertise.

They are wrong and very badly disposed. Since they comprise the greater number of known animals of the most varied characters, they should be more numerous. Hence, it has been necessary to re-constitute them and substitute others … everything which was not regarded as an insect, that is to say all invertebrate animals that have not jointed legs, were referred without exception to the class of worms … Linnæus's class of worms is a sort of chaos in which the most disparate objects are included.

Lamarck then offered his own arrangement, divided into five classes, which he had first suggested in his 1794 lectures. He adds that Cuvier joined the staff in 1795 and he learned that Cuvier also gave priority of rank to molluscs over insects, thus agreeing with Lamarck's arrangement, which had, until that time, not been favourably received.

Arrangement of Invertebrate Animals set forth in my first course.

    1. Molluscs
    2. Insects
    3. Worms
    4. Echicoderms
    5. Polyps

Lamarck continued to work on his classifications but had difficulty persuading his colleagues to embrace his ideas. He listed changes which he had instituted since 1794:

First, I changed the name of my class of echinoderms to radiarians in order to unite with them the jelly-fishes and neighbouring genera. This class … has not yet been adopted by naturalists. In my course … 1799, I established the class of crustaceans. At that time M. Cuvier … still included crustaceans with insects … it was not until six or seven years later that a few naturalists consented to adopt it. In the following year … (1800) I suggested the arachnids as a class by itself … Insects undergo metamorphosis, propagate only once in the course of their life, and have only two antennæ, two eyes with facets and six jointed legs; while the arachnids never undergo metamorphosis …

Yet this class of arachnids is still not admitted into any other work than my own.

M. Cuvier had discovered the existence of arterial and venous vessels in various animals, which used to be confused under the name worms with other animals of very different organisation. I immediately took this new fact into consideration for the improvement of my classification; and in my course in 1802, I established the class of annelids, placing them after the molluscs and before the crustaceans, as required by their organisation … I continued to place the worms after the insects, and to distinguish them from the radiarians and polyps with which they can never again be united … My class of annelids … was several years before being admitted by naturalists. For the last two years however the class has begun to gain recognition. Finally last year (1807) I established among invertebrate animals a new class – the tenth – that of infusorians.

Lamarck then listed his ten classes of invertebrate animals:

Molluscs	Insects
Cirripedes	Worms
Annelids	Radiarians
Crustaceans	Polyps
Arachnids	Infusorians

Reading Lamarck's account, two things become very clear: the rivalry between these two men and the fact that there was no clear line of demarcation between their work. Cuvier had quite clearly being working on invertebrate animals and Lamarck was probably working on vertebrates although this is harder to determine, since most of the major classification work with vertebrates had already been completed.

Adding Lamarck's ten classes to the first four of Linnæus's classes resulted in fourteen classes which Lamarck listed:

1. Mammals   )
2. Birds   ) Vertebrate animals
3. Reptiles   )
4. Fishes   )
5. Molluscs   )
6. Cirrhipedes   )
7. Annelids   )
8. Crustaceans   )
9. Arachnids   ) Invertebrate animals
10. Insects   )

11. Worms   )
12. Radiarians   )
13. Polyps   )
14. Infusorians   )

The next question to be decided was In drawing up this series, ought we to proceed from the most complex to the simplest, or from the simplest to the most complex? **Discussion of this problem would have to wait until Chapter VIII, the final chapter of Part I of Lamarck's work. First needed to be considered the** remarkable degradation of organisation which is found in traversing the natural series of animals, starting from the most perfect or the most complex towards the simplest and most imperfect ... This degradation ... so obviously and universally exists in the main groups, including even the variations, that it doubtless depends on some general law which it behoves us to discover and consequently to search for.

Chapter VI – Degradation of Organisation

LAMARCK commenced Chapter VI by making the claim that, as we progressed backwards through the series from the most complex form of animal life (mammals) to the most simple (infusoria), we would witness an increasing degradation in their organization, with a proportionate diminution of these animals' faculties ... all the special organs are progressively simplified from class to class ... they become altered, reduced and attenuated little by little ... finally they are completely and definitely extinguished before the opposite end of the chain is reached ...

The degradation of which I speak is not always gradual and regular in its progress, for often some organ disappears or changes abruptly ... again some organ disappears and re-appears several times before it is definitely extinguished ... however ... the degradation ... is nonetheless real and progressive.

Lamarck went on to say that if the factor working towards greater complexity were the only influence, then organization everywhere would no doubt have been regular, but nature had been forced to submit to the influence of the environment which had caused deviations in her progress. *Progress in complexity of organisation exhibits anomalies here and there in the general series of animals, due to the influence of environment and of acquired habits (italics in original).*

Lamarck had already referred to the life force (*aura vitalis*) which he believed to be everywhere present in Creation. As mentioned earlier, his thinking appears to have been very much in tune with that of the French Jesuit philosopher, Pierre Teilhard de Chardin (1881-1955), whose major work, *The Phenomenon of Man*, was published posthumously in 1955. Chardin also believed that there was a life force which imbued Creation, which was forever seeking new avenues for expansion. When the build up of pressure became too great to be accommodated within creation as it then was, the life force would burst forth into a new sphere, releasing the pressure. That which had burst through the barrier would become a more advanced form of life, while that which had been left behind would continue to exist in its earlier form, thus explaining why today there are still the most simple forms of life (single cell organisms multiplying by simple cell division) which have not undergone any evolution for hundreds of millions of years.

Life, wrote Chardin, was latent in inert (inorganic) matter. First to be formed were the inorganic (lifeless) spheres: the barysphere, lithosphere, hydrosphere, atmosphere and stratosphere. Then life burst through the barrier forming the biosphere and, finally, the 'noosphere', which was Chardin's term for the mental sphere of reasoning and philosophical thought, which he believed had heralded the earliest evolution within the human line.

As a priest and theologian, Chardin sought a level of evolution which distinguished humans from all other forms of life. It was an integral part of Chardin's theory that once the 'dam' had been breached, that which was left behind would never be able to progress to the next level, because the pressure had been released. No other animal, not even the Great Apes, would ever be able to enter the noosphere of reasoning thought and intellectual ability which had been attained by Man.

Lamarck, having escaped from the Jesuits, held no such philosophy. He, too, perceived a building up of 'pressure' to be the driving force behind evolution; he perceived 'breakthroughs', although he did not envision them as exclusive as the 'breakthroughs' of Chardin. Both saw organic matter, 'dead' as the result of decay, returning to an inorganic form. Many minerals, such as iron, have both an organic and an inorganic form. Animals can only absorb organic iron, after it has been taken up from the earth in its inorganic form and processed by plants. Lamarck did not know this, but he was constantly pointing out that the more deeply we looked into the workings of nature, the less clear did the boundaries seem.

It was, claimed Lamarck, the opposition of the environment to the energy of the creative force which occasionally produces ... the often curious deviations that may be observed in the progression [of deviation].

When Lamarck spoke of the environment, he assumed that the climates in various parts of the world were essentially the same now as they had always been. The concept of a Great Ice Age was still decades into the future, that of multiple Ice Ages even further away. Volcanoes and earthquakes had but local effects. The greatest change to the environment perceived by Lamarck was that created by the biota itself. Not only every species, but every *individual* of every species, impacted the environment. Every living thing feeds - ingests, digests and excretes waste matter. Every living thing respires and in so doing modifies its environment. All living forms, both plant and animal, take in water and excrete water, slightly changed, containing minerals and other chemical substances. Every living organism eventually dies and its physical body is then utilized in some way, either by being eaten or by decomposing and returning changed substances to the earth. These changes, over great lengths of time, changed the surface of the earth, and as the earth changed, so too did its inhabitants, each adapting to the changing conditions presented by the other. Change, for Lamarck, was extremely slow and it was for this reason that he did not believe that any species ever went extinct (other than at the hand of Man) since everything would have ample time to adapt.

A few centuries earlier, the Western world had struggled to adjust its thinking to accommodate the concept of an Earth which was not stable at the centre of the Universe, but constantly moving in the Heavens. Lamarck's friend and mentor, Buffon, had suggested that the Earth was not merely a few thousand years old but, rather, millions. Buffon had further suggested that the surface of the Earth had undergone changes. His ideas were supported in Scotland by James Hutton (1726-1797) and in England by Charles Lyell (1797- 1875). Not everyone was able to accept the idea that the hills and mountains, previously thought to have endured since the beginning of time, may not have existed at some time in the past, but had been formed more recently. While some were able to accept the concept of a great catastrophe forcing up the mountain ranges, many found the idea too difficult. Even more difficult was the concept that the familiar chalk hills had been formed under the sea from the skeletal remains of countless creatures which had lived æons ago. Not only had their remains been consolidated into what gave the appearance of rock, but this rock had slowly been elevated out of the sea to form the land with which we are so familiar. Lamarck's colleague, Georges Cuvier, never accepted this proposition and continued long after Lamarck's death to oppose the idea that the Earth was more than (at the most) eight thousand years old.

The ideas Lamarck was presenting were so novel, that it is worthwhile pausing for a few moments to try to put oneself back in time to the period when the Earth and its biota were believed to be the same now as they were at the time of Creation and to allow one's mind to undergo the shift in consciousness, in awareness, that Lamarck's underwent during the development of his ideas.

Lamarck assumed that life commenced in water but, right from the start, there were different environments: fresh-water, sea water, still or stagnant water, running water, the water of hot climates, of cold climates, and lastly of shallow water and very deep water. Even if the chemical structure of the first protoplasm from which life emerged was the same, life would have been confronted with different conditions and would have developed slightly differently in each of them (and each combination of them).

This is an important point, because Lamarck is about to list his fourteen classes in order, giving the impression that the one preceded the other, that, for example, insects evolved before spiders (arachnids). A casual interpretation of Lamarck's list might imply that spiders evolved from insects but from the synopsis of Lamarck's explanation, given below, it is clear that Lamarck was proposing no such thing. There was initial similarity in the basic, internal, structure within the various classes, but there was great diversity in their external characteristics, beak/teeth, digits/hooves, fur/feathers and so on. Lamarck was convinced that there were connecting links between the classes and was excited by discoveries in Australia of creatures which appeared to fill such positions, but admitted that there needed to be many more, which he was confident would yet be discovered in unexplored parts of the globe, or under the sea.

After having produced aquatic animals ... nature led them little by little to the habit of living in the air, first by the water's edge, and afterwards on all the dry parts of the globe.

Below are the fourteen classes of animal put forward by Lamarck, his definition of each class and a brief synopsis of the major points made by him in relation to each one. After Lamarck's complaints recorded in the previous chapter regarding the difficulties he experienced in having his work accepted, it is pleasing to note his comment that: The general arrangement of animals set forth above is unanimously accepted as a whole by zoologists.

Mammals

Animals with mammæ, four jointed limbs, and all the organs essential to the most perfect animals. Hair on certain parts of the body.

… the most developed intelligence … mammals alone are truly viviparous … all have a diaphragm between the chest and the abdomen; a heart with two ventricles and two auricles; red warm blood; free lungs … fœtus, although enclosed within its membranes is always in communication with its mother and develops at the expense of her substance, and in which the young feed for some time after their birth on the milk of her mammæ

… those whose limbs are adapted for grasping objects have a higher perfection than those whose limbs are adapted only for walking.

First division: unguiculate mammals; four limbs, flat or pointed claws at the end of their digits … limbs in general adapted for grasping objects or at least for hooking onto them.

Second division: ungulate mammals; … four limbs … extremity of their digits is completely invested by a rounded horn called a hoof. Their feet serve no other purpose than that of walking or running … feed exclusively on vegetable substances.

Third division: exungulate mammals; … two limbs … very short, flat and shaped like fins. Their digits are invested by skin and have no claws or horn. They have no pelvis nor hind feet;

they swallow without previous mastication ... they habitually live in the water.

Before passing on to the next class, that of birds, Lamarck made the following comments:

... there is no gradation between mammals and birds. There exists a gap to be filled, and no doubt nature has produced animals which practically fill this gap and which must form a special class if they cannot be comprised either among the mammals or among the birds.

This fact has just been realized by the recent discovery in Australia of two genera of animals, viz;

Ornithoryncus   )

Monotremes Echidna)

These animals are quadrupeds with no mammæ, with no teeth and no lips; and they have only one orifice for the genital organs, the excrements and the urine (a cloaca). The body is covered with hair or bristles. They are not mammals, for they have no mammæ and are most likely oviparous.

They are not birds; for their lungs are not pierced through and they have no limbs shaped as wings.

Finally, they are not reptiles; for their heart with only two ventricles removes them from that category.

They belong then to a special class.

Birds

Animals without mammæ, with two feet and two arms shaped like wings; the body covered with feathers.

Birds … are the only ones except the monotremes which have like mammals a heart with two ventricles and two auricles, warm blood, the cavity of the cranium completely filled by the brain, and the trunk always enclosed by ribs … they are essentially oviparous … their fœtus is enclosed in an inorganic envelope (the egg-shell) and soon ceases communication with the mother and can develop without feeding on her substance. The diaphragm … here ceases to exist, or becomes very incomplete … They breathe exclusively by lungs like the animals of the first rank; and this is not the case with any known animal after them.

Aquatic birds (like the palmipeds) … have this advantage over all other birds that their young on coming out of the egg can walk and feed … We shall then recognise that the palmipeds, waders and gallinaceans should constitute the first three orders of birds, and that the doves, passerines, birds of prey and climbers should form the last four orders of the class … their young on coming out of the egg can neither walk nor feed by themselves … the climbers are the last order of birds; moreover, they are the only ones which have two posterior digits and two anterior. This character, which they possess in common with the chameleon, appears to justify us in placing them near the reptiles.

It has been pointed out to me that farmyard chicks are able to feed themselves and walk as soon as they emerge from their eggs, as are quails.

Reptiles

Animals with only one ventricle in the heart and still possessing a pulmonary respiration though incomplete. Their skin is smooth or provided with scales.

… heart has only one ventricle … their blood is cold … they are the last animals to breathe by true lungs … in many species this organ is absent in youth and is replaced by gills … only part of the blood vessels pass through the lungs … the four limbs … begin to be lost, and indeed many of them (nearly all of the snakes) lack them altogether … there are some that are clothed in scales and others that have naked skin … although they all have a heart with one ventricle, in some there are two auricles, while in others there is only one.

Reptiles are oviparous animals (including those in which the eggs are hatched in the body of the mother) … they all have a small brain which does not fill the cavity of the cranium … the last of their order (the batrachians) … when they are first born, breathe by gills.

Lamarck explained that, although the snakes had lost their limbs, these 'appendages' were of less importance from an evolutionary point of view than the lungs. The loss of lungs in the young of the batrachians, which breathed by gills, placed this class last among the reptiles since it placed them nearer the fishes.

Fishes

Animals breathing by gills, with a smooth or scaly skin; the body provided with fins.

… the absence of a constriction between the head and the body to form a neck, and the various fins which for them take the place of limbs are the results of the influence of the dense medium they inhabit, and not of the degradation of organisation

… they have no true lungs and in its place have only gills of vascular pectinate folds arranged on both sides of the neck or head, four altogether on each side. The water by which these animals breathe goes in by the mouth … then issues through open holes on either side of the neck.

Note that this is the last time that the respired fluid enters by the animal's mouth in order to reach the organ of respiration.

These animals … have no trachea or larynx or true voice … or eyelids, etc. These organs and faculties are here lost and are not again found throughout the animal kingdom … they are … in common with the reptiles the only animals which have:

> A vertebral column;
>
> Nerves, terminating in a brain, which does not fill the cranium;
>
> A heart with one ventricle;
>
> Cold blood;
>
> Lastly, a completely internal ear.

In passing, I note that Elliot inadvertently wrote 'warm blood'. Lamarck's original words were "Le sang froid", so this was clearly an error on Elliot's part.

Fish … display oviparous reproduction … only one ventricle in the heart and cold blood; gills instead of lungs; a very small brain; the sense of touch incapable of giving knowledge of the shape of bodies; and apparently without any sense of smell, for odours are only conveyed by air.

… fishes are … divided into … bony fishes, which are the most perfect, and cartilaginous fishes, which are the least perfect … among the cartilaginous fishes the softness and cartilaginous condition of their parts … indicate that it is among them that the skeleton ends or rather that nature has sketched its first rudiments.

Observations on the Vertebrates

The vertebrate animals … appear all to be formed on a common plan of organisation … this plan while approaching perfection has undergone numerous modifications, some of them very large, through the influence of the environment … and of the habits which each race has been forced to contract by the conditions in which it is placed.

… nature's work has often been modified, thwarted and even reversed by the influence exercised by very different and indeed conflicting conditions of life …

Annihilation of the Vertebral Column

On reaching this point in the animal scale the

vertebral column becomes entirely annihilated … the supports for muscular activity will no longer reside in any internal parts.

Moreover, none of the invertebrate animals breathes by cellular lungs; none of them has any voice nor consequently any organ for this faculty; finally they appear to be devoid of true blood … we also lose here the iris … such of the invertebrates as have eyes have no definite irises.

Kidneys in the same way are only found among vertebrates … henceforth there is no more spinal cord, no more sympathetic nerve.

… among vertebrates … all the essential organs are isolated or have each an isolated seat in as many special places.

Invertebrate Animals

On reaching the invertebrate animals … we shall not meet with a single system of organisation progressively perfected, but with various quite distinct systems, each one taking its start at the point where each organ of highest importance began to exist.

Molluscs

Soft unjointed animals which breathe by gills and have a mantle. No ganglionic longitudinal cord; no spinal cord.

… molluscs … have to be placed a stage lower than the fishes since they have no vertebral column … They breathe by gills … They all have a brain; nerves without nodes, that is to say without a row of ganglia stretching down a longitudinal cord. They have arteries and veins and one or many single chambered hearts. They are the only known animals which, although possessing a nervous system, have neither a spinal cord, nor a ganglionic longitudinal cord.

… The respiratory organ of these animals has imperceptibly become accustomed to air … this habit of breathing air with gills became a necessity to many molluscs which acquired it.

… we may distinguish two kinds of gills. The first kind consists of networks of vessels running through the skin of an internal cavity which is not protruded and can only breathe air: these may be called aerial gills. The second kind are organs nearly always protruded … forming fringes or pectinate lamellæ or edgings, etc.; these can only achieve respiration by means of the contact of fluid water, and may be called aquatic gills.

… A lung is essentially a peculiar spongy mass composed of more or less numerous cells into which air is always entering in nature. The entrance is effected through the animal's mouth and thence by a more or less cartilaginous canal called the trachea … The cells and bronchi are alternately filled and emptied of air by successive swellings and shrinkings of the cavity of the body containing the mass; so that distinct alternate inspirations and expirations are characteristic of a lung … the bronchial cavity of certain molluscs, which is quite peculiar, exhibits no alternate swelling and shrinking, never has a trachea, or bronchi and … the inspired fluid never enters by the animal's mouth … After the reptiles no animal has a lung; nor therefore a voice.

There are two orders of mollusc … the first of these orders (cephalic molluscs) have a very distinct head, eyes, jaws or a proboscis and reproduce by copulation. All the molluscs of the second order (acephalic molluscs) are destitute of a head, eyes, jaws, proboscis; and they never copulate for the purpose of reproduction … We have here … one of those deviations in the progress of perfection of organisation that are produced by environment and consequently by causes foreign to those which make for a gradual increase of complexity…

In conformity with the law of nature which requires that every organ permanently disused should imperceptibly deteriorate, become reduced and finally disappear; the head, eyes, jaws, etc., have in fact become extinct in the acephalic molluscs …

In the invertebrates nature no longer finds in the internal parts any support for muscular movement; she has therefore supplied the molluscs with a mantle for that purpose … in the cephalic molluscs, where there is more locomotion than in those which have no head, the mantle is closer, thicker and stronger; and among the cephalic molluscs, those which are naked (without shells) have in addition a cuirass in their mantle which is stronger than the mantle itself and greatly facilitates the

locomotion and contraction of the animal
(slugs).

Cirrhipedes

*Animals without eyes which breathe by gills and
have a mouth and jointed arms with a horny skin*

The cirrhipedes, of which only four genera are
yet known, should be considered a special class
… their nervous system is characterized like91
the animals of the following three classes by a
ganglionic longitudinal cord. They have …
jointed arms with a horny skin and several
pairs of transverse jaws … their fluids move by
a true circulation with arteries and veins.
These animals are fixed on marine bodies and
consequently carry out no locomotion; their
principal movements are those of their arms …
their skin is coriaceous [leathery] and almost
horny like that of crustaceans and insects.

Annelids

*Animals with elongated annulated [ring-shaped]
bodies without jointed legs, breathing by gills
and having a circulatory system and a
ganglionic longitudinal cord*

… no annelid has a mantle … they have no
jointed legs … they are inferior to the
molluscs in that they have a ganglionic
longitudinal cord … Annelids owe their
elongated form to their habits of life, for
they either live buried in damp earth or in mud
or actually in the water mostly in tubes of

various sorts which they enter and leave at will. [They have] a very small brain, a ganglionic longitudinal cord, arteries and veins in which circulates blood that is usually coloured red; they breathe by gills.

Crustaceans

Animals with a jointed body and limbs, crustaceous skin, a circulatory system, and breathing by gills

We now enter the long series of animals, whose bodies and limbs are jointed ...The solid or hard parts ... are all on the exterior ... held by Linnæus and subsequently as forming only a single class ... insects. The crustaceans indeed have a heart, arteries and veins; a transparent and almost colourless circulating fluid, and they all breathe by true gills ... Their nervous system ... consists of a very small brain and a ganglionic longitudinal cord ... It is in the crustaceans that the last traces of an organ of hearing have been identified; after them, it is no more found in any animal. Here ends the existence of a true circulatory system.

Arachnids

Animals breathing by limited tracheæ, undergoing no metamorphosis, and having throughout their lives jointed legs and eyes in their head

... the arachnids furnish us with the first example of a respiratory organ lower than gills, – one never met with in animals which have a heart, arteries and veins. Arachnids ... breathe only by stigmata and aircarrying tracheæ ... limited to a small number of sacs ... Arachnids:

– never undergo metamorphosis ... they have eyes on their head and jointed legs throughout their lives ...

– in arachnids of the first order (pedipalp-arachnids) we begin to see the outline of a circulatory system ...

– tracheæ are limited to a small number of sacs ... not ...extending throughout the animal's body ...

– arachnids procreate several times in the course of the life; a faculty which the insects do not posses

In a footnote, Lamarck quoted Cuvier as having observed that in arachnids the heart may be easily observed beating through the skin of the abdomen in species not hairy, from the sides of which could be seen issuing two or three vessels. It would seem that Cuvier had been studying, and publishing material about arachnids, even though he held the Chair of vertebrate animals. Was this a breach of professional etiquette? It certainly would have done nothing to heal the ever-widening divide between these two 'colleagues'.

Insects

Animals which undergo metamorphoses, and have in the perfect state two eyes and two antennæ in their head, six jointed legs and two tracheæ which extend throughout the body

... insects ... have no arteries or veins ... breathe

by air-carrying tracheæ not limited to special parts … born in a state less perfect than that in which they reproduce … subsequently undergo metamorphoses. In their perfect state all insects without exception have six jointed legs, two antennæ and two eyes in their head, and most of them also have wings

… only procreate once in the course of their life … organs essential to maintenance of life are almost equally distributed … throughout their bodies instead of being isolated in special places.

… the first three orders, (Coleoptera, Orthoptera, Neuroptera) have mandibles and maxillæ in their mouths … the fourth order (Hymenoptera) begin to possess a sort of proboscis … the last four orders (Lepidoptera, Hemiptera, Diptera and Aptera) have really nothing more than a proboscis … paired maxillæ are nowhere found again in the animal kingdom, after the insects of the first three orders … insects of the six first orders have four [wings], all of which or only two serve for flight. Those of the seventh and eighth have only two wings or else they are quite aborted. The larvæ of the insects of the two last orders have no legs and are like worms … the insects are the last animals which have a quite distinct sexual reproduction

… these are the only invertebrate animals which launch themselves into the air

… it appears there is a rather large gap … remaining to be filled by animals not yet observed; for in this part of the series several organs essential to the most perfect animals suddenly drop out

… the nervous system … completely disappears … there is no longer a separate nucleus for sensations, but a multitude of small nuclei scattered throughout the length of the animal's body … among insects, the important system of feeling comes to an end … the source whence muscular action derives its power and without which sexual reproduction apparently could not exist

… the organ of sight, so useful to the most perfect animals, is [hereafter] entirely extinguished … the head altogether ceases to exist.

Worms

Animals with soft elongated bodies, without head, eyes or jointed legs, and no longitudinal cord or circulatory system

… worms have no vessels of circulation … undergo no metamorphosis and are all destitute of head, eyes and jointed legs … we no longer find … bilateral symmetry

… several worms still appear to breathe like insects by tracheæ of which the external openings are kinds of stigmata … these limited or imperfect tracheæ are water-carrying and not air-carrying like those of insects … these animals never live in the open air, but are continuously in the water or bathed by fluids which contain water

… As no organ for fertilisation is distinguished in them, I suppose that sexual reproduction does not occur in these animals … Objects … found in some of them and supposed to be ovaries … appear to be merely clusters of reproductive corpuscles which do not require fertilization … it is therefore probable that worms are internally gemmiparous.

Radiarians

Animals with regenerating bodies, destitute of a head, eyes or jointed legs; with a mouth on the inferior surface and a radiating arrangement of the parts, both internal and external

… radiating arrangement about a centre or axis … organs apparently intended for respiration (tubes or kinds of water-bearing tracheæ)

… special organs for reproduction, such as kinds of ovaries of various shapes … their mouth is always on the inferior surface … no head, eyes or jointed legs, circulatory system or perhaps nerves

… fibres may still be distinguished; but … it does not follow that because a living being has distinguishable fibres, it must therefore have muscles; I hold that where there are no nerves, there is no muscular system …. Animals without nerve fibres … possess the faculty by mere irritability of producing movements … the muscular system has ceased to exist in radiarians … there seems to be no sexual reproduction … I regard … internal gemmules …

nature's preliminary step towards sexual reproductions. … the intestinal canal no longer has two exits … the body becomes completely gelatinous.

Polyps

Animals with sub-gelatinous and regenerating bodies, with no special organs but an alimentary canal with only one opening. Terminal mouth supplied with radiating tentacles or a ciliated and rotary organ

… gemmiparous animals with homogenous bodies, usually gelantinous, and with very regenerative parts not displaying the radiating shape … except in radiating tentacles around their mouth and having no special organs but an intestinal canal which has only one opening … no brain, longitudinal cord, nerves, special respiratory organs, vessels for the circulation of fluids nor ovary for reproduction … the parts are nothing more than merely irritable … they are devoid of feeling and hence of every kind of sensation … only move by external stimuli foreign to themselves … movements … carried out without any act of will … polyp swallows [automatically] … without making any distinction as to … suitability [for digestion]… brings up debris … superior mouth, while the mouth of the radiarians is otherwise situated …

Infusorians

Infinitely small animals with gelatinous, transparent, homogenous and very contractile bodies; with no distinct special organ internally, but often oviform gemmules; and having externally no radiating tentacles nor radiating organs

… the last of these genera shows us in some degree the limit in animality … very tiny gelantinous, transparent, contractile and homogenous bodies, consisting of cellular tissue, with very slight cohesion and yet irritable throughout … feed by absorption and continual imbibition

… It is exclusively among animals of this class that nature appears to carry out direct or spontaneous generation … these fragile animals perish during the reduction of temperature in bad seasons … They are found just the same in all parts of the world, but only in conditions suitable for their existence

… even before leaving the division of vertebrates, we already witness great changes in the perfection of organs; while some even disappear altogether, such as the urinary bladder, the diaphragm, the organ of voice, the eyelids, lungs … but it is in the division of the invertebrates that the extinction takes place of the heart, brain, gills, conglomerate glands, vessels for circulation, the organs of hearing and sight, that of sexual reproduction and even that of feeling, as also of movement

… Thus on traversing the chain of animals from the most perfect to the most imperfect, and on examining in turn the various systems of organisation distinguished in the course of this chain, the degradation of organisation and of each organ up to their complete disappearance is seen to be a positive fact which we have now verified.

Comments

The first point to be noted is Lamarck's tentative inclusion of a fifth class of vertebrate animal – the monotremes. Why did Lamarck not complete the inclusion? Was it a professional courtesy to his colleague, George Curvier, who held the Chair of Vertebrate studies? Lamarck had no compunction about amending Linnæus' classification of invertebrates and I can think of no other valid reason for his not having amended the vertebrate classification also.

The second point to note is that in this chapter Lamarck addressed the first of the two factors which he held shaped the progress of evolution – that of the 'inner' force, the *aura vitalis*. In the next chapter he addressed the second factor, the environment. All of the changes, which he counted as 'degradations', since he was following their course backwards from humans to infusoria, had been 'influenced' by this inner force and all of these changes referred to 'inner' organs essential to the life of the animal at that stage of evolution. Lamarck was not, at this time, concerned with feathers and fur, claws and paws. How many legs an animal had, or even if it had any at all, was influenced by the environment in which the animal lived and was, therefore, peripheral.

Imagine a tree. As the tree begins to grow, it puts forth branches. Think of these as the classes, initially one or two, first the infusoria and then the polyps. Lamarck's comments about the different 'waters' in which these creatures might find themselves, fresh, salty, deep, shallow, etc., show that Lamarck envisioned that they, according to the environment in which they found themselves, might develop somewhat differently, that is, that there might be more than one 'family' or 'genera', which, in the analogy being offered, would manifest as smaller branches or twigs growing from the main branch, which, itself, emanated from the main trunk of the tree. As the tree grew, more branches would develop. These would not grow from the twigs, or smaller branches, of the previous main branch, but would emanate from the main trunk.

At the beginning of his section on invertebrates, before discussing the molluscs, Lamarck stated of the various quite distinct systems that we were going to meet that each one would take "*its start at the point where each organ of highest importance began to exist*" *(emphasis added)*. The point where each organ had "begun to exist" was at the trunk, where the previous branch (class) had branched out. The trunk had continued to grow and, subsequently, put forth further branches. Each branch (class) subdivided into families, genera, species, varieties, but none of these families, genera, species, varieties, had any contact *whatsoever* with the branches and twigs (families, genera, species and varieties) from the branch below. Lamarck made it quite clear, for example, that he did not consider that there was any direct connection between the insects and the arachnids.

The life force *(aura vitalis)*, which in this analogy is represented by the sap, having matured, or changed in some way as it moved upwards, would never thereafter give rise to branches devoid of the latest *essential* organ. The higher classes would express these higher organs, but they may or not be possessed of similar peripheral features, such as hair or legs, which may have developed in the lower branches (classes) according to the requirements of their environment.

It was a misunderstanding of this very important point which led Elliot (xxxiv-xxv) to think that Lamarck had taught that "man's ancestors include every existing species of animal. Not only had he bird, reptile and fish ancestors, but also arachnid, insect, worm, starfish, etc., ancestors. He passed through the stage of being a scorpion and a spider. He traversed in turn every known species of insect. He was a tapeworm, a sea-anemone, a polyp and an amoeba." Not surprisingly, Elliot declared this doctrine to be "totally absurd". It would have been, if that had, indeed, been Lamarck's teaching, every bit as absurd as it would have been if Lamarck had taught that a main branch of a tree grows from the twig of the lower branch. Everything grows from the main stem, trunk, and draws its sap therefrom.

Chapter VII – The Influence of the Environment

IN CHAPTER VI, Lamarck concentrated on the effect of an internal life-force on the evolution of all living organisms. No species emerging from any branch of the 'tree' could develop any essential organ beyond that of any other species emerging from that branch, or from any previous branch. There was a limit to the possibilities of their 'organisation'. This energy could be considered 'general' inasmuch as it controlled the general organization of the class.

In Chapter VII, Lamarck looked at the second 'force' which he perceived as influencing evolution – the environment. The environment itself is passive. It does nothing to alter any living thing. It is not the environment, *per se*, which influences evolution, but the reaction of the organism to the environment.

... *the environment affects the shape and organisation of animals* ... when the environment becomes very different, it produces in course of time corresponding modifications in the shape and organisation of animals ... whatever the environment may do, it does not work any direct modification whatever in the shape and organisation of animals ... great alterations in the environment of animals lead to great alterations in their needs, and these alterations in their needs necessarily lead to others in their activities. *(Italics in original)*

A change in the needs of an animal might lead that animal to make greater use of a certain part, resulting in the further development of that part, or, alternatively, it might lead the

93

animal to use a certain part less, resulting in that part's diminution or even total disappearance. The general organization of all vertebrates dictated four limbs. No vertebrate, in pursuit of extra speed, or greater stability, had ever developed six legs. However, as the result of disuse, some vertebrates had lost some, or all, of their limbs, some had greatly decreased their size, others had converted two of their limbs to a use other than walking/running. Legs were a hindrance to snakes slithering through small spaces and so these appendages disappeared. Eyes were not a hindrance to moles living underground, but they were surplus to requirements and so these, too, gradually disappeared, although the original 'plan' was still evident.

Lamarck drew on changes which had occurred under domestication in both plants and animals to support his argument that changes in environment precipitated changes in form:

Where in nature do we find our cabbages, lettuces, etc., in the same state as in our kitchen gardens? and is it not the same with regard to many animals which have been altered or greatly modified by domestication? … our domestic ducks and geese are of the same type as wild ducks and geese; but ours have lost the power of rising into high regions of the air and flying across large tracts of country … Where in natural conditions do we find that multitude of races of dogs which now actually exist, owing to the domestication to which we have reduced them?

… in the case of animals, extensive alterations in their customary environment produce corresponding alterations in their parts; but here the transformations take place much more slowly than in the case of plants … but what is not known as well and indeed what is not generally believed, is that every locality itself changes in time as to its exposure, climate, character and quality, although with such extreme slowness, according to our notions, that we ascribe to it complete stability … altered localities involve a corresponding alteration in the environment of the living bodies that dwell there and this again brings a new influence to bear on these same bodies.

Lamarck was here drawing upon the thoughts and ideas of his mentor, Le Comte du Buffon, in the same way as half a century later Charles Darwin was to draw upon the ideas of his mentor, Sir Charles Lyell, whose three volume work, *Principles of Geology* (1830-1833), so influenced the thinking of Victorian England. Even today some people are not prepared to accept the great age of the Earth, still holding that it was created less than 10,000 years ago, mostly fundamental Christians and orthodox Jews and Muslims, who number in the millions. The change of thinking necessary to adopt the position supported by Buffon and his pupil, Lamarck, was very great and should not now be underestimated simply because many of us are educated in this concept from a very young age. First it was necessary to accept the reality of great changes to the surface of the Earth which had taken place over great expanses of time. Only then was it possible to embrace the concept of corresponding changes in plant and animal life, including that of humans.

Lamarck then outlined three basic principles:

1) Every fairly considerable and permanent alteration in the environment of any race of animals works a real alteration in the needs of that race;

(2) Every change in the needs of animals necessitates new activities101 on their part for the satisfaction of those needs, and hence new habits;

(3) Every new need, necessitating new
activities for its satisfaction, requires the
animal, either to make more frequent use of
some of the parts which it previously used
less, and thus greatly to develop and enlarge
them; or else to make use of entirely new parts
to which the needs have imperceptibly given
birth by efforts of its inner feeling;

... the infinitely diversified but slowly
changing environment in which the animals of
each race have successively been placed, has
involved each of them in new needs and
corresponding alterations in their habits.

The theory of 'use/disuse' leading to the
development/deterioration of related parts is largely
uncontroversial, being a fundamental tenet of Darwinism as well
as of Lamarckism. The most significant part of the above
principles is the last: "to make use of entirely new
parts to which the needs have imperceptibly
given birth by efforts of its inner feeling".
One of the criticisms of Darwin's theory was that it accounted for
variation in existing features but not for the origin of any feature,
let alone species, despite the title of the book! Lamarck's
'explanation' was philosophical, not scientific, if for no other
reason than that the science of reproduction was not at that time
understood.

The gradual improvement in the efficiency of microscopes had
allowed sperm to be identified within the seminal fluid, but not in
any great detail. At that time, Graff had identified what he
believed was the female egg, but which later was known to be
the Graffian follicle. Some had suggested that offspring were
formed entirely from the sperm, or animalcule as it was then
known, and that the female merely provided 'nutrition' for the
developing foetus. Others believed the opposite, that the embryo

developed from the female egg and that the male seminal fluid, including its numerous animalcules, provided the 'nutrition'. Fertilization had not yet been observed under the microscope.

The great French mathematician and scientist, Pierre Louis Moreau de Maupertuis (1698-1759), who died exactly one hundred years before the publication of *Darwin's On the Origin of Species* and fifty years before the publication of Lamarck's *Philosophie Zoologique*, was already pondering many of the same problems which were so to fascinate them. In *The Earthly Venus*, published in 1753, Maupertuis rejected the idea that 'nutritive substances' played any role in the characteristics of the developing fœtus, arguing that, if nutritive material could so influence the growing embryo, why did not the human infant resemble the fruit, vegetables and other nutritive substances eaten by the mother during her pregnancy?

Maupertuis believed that both the egg and the sperm (animalcules) contained particles responsible for the formation of parts of the body. Maupertuis believed that at conception these particles would gravitate towards, and join with, their most closely related particle. Particles from either parent would have an equal chance of forming any particular part of the embryo. If the process worked perfectly, a 'perfect' child would result. If some particle were missing or became damaged, a child deformed by deficiency would be born. If some particle in excess of that required somehow managed to attach itself, then extra body parts would result, even to the extent of producing conjoined twins, infants with two heads or with one head and two bodies.

Maupertuis studied the family history of two families in Paris, some of whose members exhibited polydactily, the presence of an extra digit on at least one of their hands or feet. He was able to show that this family trait could be passed on by either the male or the female. Thus, by the time Lamarck was writing, it was accepted that both parents had an equal chance of contributing towards the formation of their mutual offspring. It was not yet known precisely how this occurred and Lamarck, as he had stated in his Preface, declined to speculate upon that which was not certain. Darwin, who loved 'philosophising', was not so hesitant. His theory of pangenesis, given in his *Descent of Man* published in 1871, proposed that there were 'gemmules' present in the blood which 'picked up' on any changes which had occurred in the body of their host, brought that information back to the reproductive organs, thus enabling the new characteristic to be incorporated into future offspring.

Pangenesis was essential to Darwin's theory because he saw change as arising from the individual. In this, his theory differed from that of Lamarck who, as we shall see as we progress through the remainder of this chapter, held that evolution took place as the result of changed conditions experienced by specific *populations* over an extended period of time.

There is a similarity between the thinking of Lamarck and that of Sheldrake (1988), who proposed the existence of an hitherto unknown energy he called the morphic field. The energy fields of magnetism and gravity had long been acknowledged and the latter had had a profound influence on evolution. Every living thing, from the blade of grass pushing upwards through the soil to the eagle soaring in the sky, must adapt to the requirements of gravity if it is to exist as a living entity. Scientists knew that once a new chemical had crystallized for the first time, scientists in other laboratories in other parts of the world would find it easier to form this crystal on subsequent occasions. Sheldrake proposed that a 'morphic field' had been created by the new crystal which permeated the ether in some way such that other similar substances would resonate therewith, making subsequent crystallization easier.

Sheldrake's 'morphic field', so called because it caused substances to change their shape, i.e. morph, appears to have been a passive force. It was simply there, composed of the energy fields of both inert matter, such as rocks, and living matter. Everything contributed to the morphic field of its environment simply by being where it was. Fields fluctuated, seasonally as animals came and went in a particular environment, over longer periods of time as landscapes changed and their biota changed with it, but all contributed to the changing field and *all were affected by that change*. Change was automatic according to the force field created at the time.

Sheldrake extended this principle in relation to higher life forms, especially humans. He suggested that there were morphic fields for behavior and memory which would influence the thoughts and behaviors of a peer group; he further suggested that we may resonate with the thoughts and behaviors of people similar to ourselves who had lived in the past. This was somewhat similar to Carl Jung's (1963) concept of Universal Consciousness and Richard Dawkins' (1976) theory of memes.

Lamarck's concept was slightly different. He, too, seems to have envisaged every species of plant or animal as having been influenced by some type of energy field created by that population in response to the environment in which it found itself. Although Lamarck entitled his chapter "The Influence of the Environment", it is quite clear from reading the chapter that the influence came from the biota's reaction to the environment in which it found itself. His use of the term 'inner feeling' seems to imply some sort of sub-conscious reaction, a reaction common to all the population of similarly formed individuals who found themselves in the same environment and all of which reacted in the same manner. Change only took place when these 'inner feelings' were maintained over long periods of time, during which they, in some unknown way, influenced the future shape and form of the evolving species.

Lamarck then stated his two laws, which he gave in italics.

First Law

In every animal which has not passed the limits of its development, a more frequent and continuous use of any organ gradually strengthens, develops and enlarges that organ, and gives it a power proportional to the length of time it has been so used; while the permanent disuse of any organ imperceptibly weakens and deteriorates it, and progressively diminishes its functional capacity, until it finally disappears.

Second Law

All the acquisitions or losses wrought by nature on individuals, through the influence of the environment in which their race has long been placed, and hence through the influence of the predominant use or permanent disuse of any organ; all these are preserved by reproduction in the new individuals which arise, provided that the acquired modifications are common in both sexes, or at least to the individuals which produce the young.

The first law could be applied to the development of inner organs, of which Lamarck wrote in Chapter VI. There was a limit beyond which development could not take place. However, Lamarck was here referring to peripheral evolution, changes which take place in non-essential organs, which are extremely varied, particularly among insects. Unlike Darwin, who always insisted that there was no limit to the variation which could take place in nature, Lamarck acknowledged that there were limits and scientists today would agree with him. There is a limit to the height and distance to which a heart can pump blood; there is a limit to the weight of a body which can be lifted into the air. Apart from this, Lamarck's first law was embraced by Darwin, although he always denied that he had been influenced by Lamarck's work.

The second law at first seems to imply that, as claimed by Darwin, change occurs in individuals and is passed on by them. However, the proviso at the end of the law makes clear that Lamarck only saw change being passed on to the next generation when it had become sufficiently established to occur, at the very least, in two individuals – those who were mating. The chances of only two animals experiencing some form of change, one male and one female, and of those two being the two who happened to mate, are very small. It would seem that Lamarck only saw change being passed on by reproduction after it had become established in the group, although quite how it became established in the group was not clear. The only explanation was the working of the 'inner feeling' to which Lamarck had already referred.

In an effort to explain his position, Lamarck pointed out that, until that time, naturalists held that the structure of a part was dictated by its function. He held the contrary view that function determined structure. Function was determined by the animal's needs in the environment in which it found itself. It was these needs, which changed as the environment changed, which gave birth to them [parts] when they did not exist. If structure came first, there would be no change. If the structure became unable to perform a required function, the animal would not be able to continue to exist. Only if function determined structure would change (evolution) come about.

It is not the organs, that is to say, the nature and shape of the parts of an animal's body, that have given rise to its special habits and faculties, but it is, on the contrary, its habits, mode of life and environment that have in course of time controlled the shape of its body, the number and state of its organs and, lastly, the faculties which it possesses.

Lamarck then gives several pages of examples, including that of the giraffe reaching for leaves, which is so frequently cited to illustrate Lamarck's theory: it is obliged to browse on

the leaves of trees and to make constant
efforts to reach them. From this habit long
maintained in all its race, it has resulted
that the animal's fore-legs have become longer
than its hind-legs, and that the neck is
lengthened to such a degree that the giraffe,
without standing on its hind legs, attains a
height of six metres (nearly 20 feet).

In this paragraph, Lamarck again stresses that change comes about as the result of a habit, long maintained in all the members of the race in question. It does not result from a member fortuitously having been born with a neck slightly longer than normal having an advantage, being 'selected' by nature, and giving birth to a number of offspring similarly endowed with a slightly longer neck. Change takes place in the whole population and becomes established very gradually.

In the previous paragraph, Lamarck had been discussing the fights which took place between ruminant males and commented that:
In the frequent fits of anger to which the
males especially are subject, the efforts of
their inner feeling cause the fluids to flow
more strongly towards that part of their head ...
Note, again, Lamarck's reference to 'their inner feeling', a feeling which is common to all the males of the species and which directs the flow of vital energy (fluids) to a certain area, resulting, in this case, in the development of horns.

Elliot, in his Introduction, complained that Lamarck was repetitive and that may be true, but Lamarck was endeavouring to stress important points which, unless fully grasped, would lead to the student/reader not understanding his theory. He gave a few more examples after the ruminants and giraffes and then restated his position:

... efforts in a given direction, when they are
long sustained or habitually made by certain

parts of a living body, for the satisfaction of needs established by nature or environment, cause an enlargement of these parts and the acquisition of a size and shape that they would never have obtained, if these efforts had not become the normal activities of the animals exerting them.

After all these reiterations, Lamarck would hope that he had made it quite clear that he did not see evolution as having come about as the result of a fortuitous change in a single individual, subsequently passed on by inheritance. Change came about as the result of an inner feeling or drive present in all members of the population long sustained. Referring once again to domesticated animals, he continued:

I can, in short, cite a multitude of instances among ourselves, which bear witness to the differences that accrue to us from the use or disuse of any of our organs, although these differences are not preserved in the new individuals which arise by reproduction; for if they were their effects would be far greater.

... every change that is wrought in an organ through a habit of frequently using it, is subsequently preserved by reproduction, if it is common to the individuals who unite together by fertilisation for the propagation of their species ... Furthermore, in reproductive unions, the crossing of individuals who have different qualities or structures is necessarily opposed to the permanent propagation of these qualities and structures. Hence it is in man, who is exposed to so great a diversity of environment, the accidental qualities or defects which he acquires are not preserved and propagated by reproduction.

It is quite clear from these passages that Lamarck did not see evolution as having resulted from a change, which occurred either at birth or during the life of an individual, being passed on to subsequent generations by inheritance. Change was only inherited after the need for that change had been experienced by the population over a considerable period of time.

This is the first mention by Lamarck of humans in relation to evolution and he makes it with the implicit assumption that humans have been subject to the same evolutionary forces as all other forms of life. This was something Darwin was reluctant to do. Some have suggested that Darwin's reluctance stemmed from respect for the opinions of his wife, Emma, who was very religious. Christian teaching, based on the Bible, and especially the story of Creation as recorded in the Old Testament, was that Man was a special creation, with a soul and the knowledge of good and evil, attributes with which animals were not endowed. Lamarck, having abandoned Christian beliefs, had no such reservations.

In one sense, it would be true to say that both Lamarck and Darwin believed in the inheritance of acquired characteristics, but their teachings were quite different. Darwin held that changes which occurred in an individual could be passed on by inheritance and had the potential to initiate a new line of evolution. Lamarck clearly denied that change in a single individual could have any such effect.

It is difficult to understand quite how Lamarck envisioned evolution as having taken place if change could only be passed on by inheritance once it had become common to an interbreeding population. Thinking as we do today of genes and DNA, it is hard to imagine this process in action. However, genetics does support Lamarck's claim that the crossing of individuals with different characteristics is opposed to the permanent establishment of a new characteristic. It was well known that domestic breeders of both plants and animals needed to isolate the individuals from whom they wished to breed a new characteristic because, if these individuals were allowed to breed within their larger population, the desired characteristic would 'breed out'. It is now known that most mutations observed in the laboratory occur as recessives, meaning that they are not expressed in the phenotype and are not, therefore, available for 'selection', natural or otherwise. In the laboratory, many individuals are subjected to the same conditions, such as irradiation, causing many to mutate in a similar manner. With large populations, such as fruit fly, there is a reasonable chance that two such similarly mutated individuals will mate and a deviation will be expressed in the next generation. Quite how this would take place in the wild has not been fully explained, even today. Those who accept evolution are forced simply to accept that it has, as was Lamarck.

There were clear gaps in Lamarck's knowledge and understanding of the evolutionary process, but his lack of knowledge and understanding was that of his time, not an individual failing. He had clearly studied long and thought deeply and presented to the world a new vision, one that no one had ever before conceived.

Lamarck concluded this chapter by saying that we were faced with two options: Conclusion adopted hitherto: Nature (or her Author) in creating animals, foresaw all the possible kinds of environment in which they would have to live, and endowed each species with a fixed organisation and with a definite and invariable shape, which compel each species to live in the places and climates where they actually find them, and there to maintain the habits which we know in them.

My individual conclusion: Nature has produced all the species of animals in succession, beginning with the most imperfect or simplest, and ending her work with the most perfect, so as to create gradually increasing complexity in their organisation; these animals have spread at large throughout all the habitable regions of the globe, and every species has derived from its environment the habits that we find in it and the structural modifications which observation has shown us …

Can there be any more important conclusions in the range of natural history, or any to which more attention should be paid than that which I have just set forth?

Chapter VIII – The Actual Order of Nature

THUS far, Lamarck had considered the natural order of animals in the manner usually adopted – from the most complex (humans) downwards to the least complex (single-celled infusoria). This may have been acceptable if it was assumed that all animals were created at the one time, but was clearly the reverse of the natural order, if it was assumed that animals had evolved from the singlecelled infusoria and progressed to the most complex organisation ... the only direct production that occurs ... is in the case of the simplest organized bodies ... these she [Nature] continues to produce ... in the same way at favourable times and places. Lamarck returns later to the subject of spontaneous generation, so consideration of this controversial topic will here be deferred.

Lamarck now returned to the fourteen classes which he had previously outlined, but considered them now in their natural order, this time numbering the classes, rather than ranking them, so that the infusoria were Class I and Mammals Class XIV. Lamarck gave six groupings of stages of evolution:

Stage 1	Infusorians	)	No nerves; no vessels; no specialized
	Polyps	)	internal organ except for digestion
Stage 2	Radiarians	)	No ganglionic longitudinal cord; no
	Worms	)	vessels for circulation; a few
			internal organs in addition to those
			of digestion
Stage 3	Insects	)	Nerves terminating in a ganglionic
	Arachnids	)	longitudinal cord; respiration by
			air-carrying tracheæ; circulation
			absent or imperfect
Stage 4	Crustaceans	)	Nerves terminating in a brain or a

Annelids	)	ganglionic longitudinal cord;
Cirrhipedes	)	respiration by gills; arteries
Molluscs	)	and veins for circulation
Stage 5 Fishes	)	Nerves terminating in a brain which
Reptiles	)	is far from filling the cranial cavity; heart with one ventricle and the blood cold
Stage 6 Birds	)	Nerves terminating in a brain which
Mammals	)	fills the cranial cavity; heart with two ventricles and the blood warm

This is an abbreviated version of Lamarck's table, which also included the organizational differences between each member of each group given in Chapter VI and which ran to thirty-five pages.

Lamarck then worked his way forward through the fourteen classes, making observations on each and giving the orders in each class. Some classes, such as infusorians, only had two orders (naked infusorians; appendiculate infusorians), while others, such as birds, had many (climbers, birds of prey, passerines, columbæ, gallinaceans, waders, palmipeds). For each order, Lamarck listed the families found therein. He finished with the mammals and their four orders, exungulate mammals which included whales and dolphins, amphibian mammals included seals and dugongs, ungulate mammals which included cattle and giraffe and unguiculate mammals which included seven families from ant-eater to the monkeys and apes, which last group he termed 'quadrumanus'. After the birds and before the mammals, Lamarck separately listed the monotremes with their two classes, ornithorhynchus and echidna. He concluded this section by remarking:

Naturalists who have considered man exclusively according to the affinities of his organisation, have formed a special genus for him with six known varieties, thus making him a

separate family which they have described in
the following manner:

Bimana

Mammals with differentiated unguiculate limbs;
with three kinds of teeth and opposable thumbs
on the hands only.

Man

(Varieties)

Caucasian

Hyperborean

Mongolian

American

Malayan

Ethiopian or Negro

This family has received the name Bimana,
because in man it is only the hands that have a
separate thumb opposite to his fingers, while
in the Quadrumanus the hands and feet have the
same character as regards the thumb.

Lamarck was here being uncharacteristically cautious. He does
not cite a specific source for the above characterization; it would
seem to be sufficiently well-known for this not to be necessary.
However, it is clearly not of his making, since Lamarck's
summaries never included peripheral, and therefore unimportant,
characteristics such as teeth or limbs.

Lamarck then asked: what if force of circumstances
caused some race of quadrumanous animals to
leave the trees, to use their feet only for
walking on the ground, if they needed to stand
upright in order to see their surroundings, and
what if this race stopped using their jaws for

biting or grasping, and as a result of these changes having gained supremacy over other races, might not this race have become dominant; have changed its habits as a result of the absolute sway exercised over the others, and of its new wants; have progressively acquired modifications in its organisation, and many new faculties; have kept back the most perfect of the other races to the condition that they had reached; and have wrought very striking distinctions between those last and themselves.

Despite his somewhat cautious start, Lamarck is here quite distinctly claiming that the human genus had descended from a race of apes, or orangs as they were then called. Fifty years later, Charles Darwin drew back from making the same claim. It appeared to be the logical extrapolation of this theory but he did not mention human evolution in *On the Origin of Species*. His good friend, Thomas Huxley, who, four years later, in 1863, published *Man's Place in Nature*, undertook this task, arguing forcefully for human descent from either the orang or the chimpanzee. The shy gorilla had not yet been seen by Europeans; only the Simia troglodytes, as chimpanzees (pan troglodytes) were then called, and the orang-utang were known. Lamarck considered the Simia troglodytes more perfect than the orang-utang. He noted that, despite their ability to stand upright in some degree, the quadramanous would drop to all fours when speed was necessary. Lamarck also noted, that despite the erect position being easy for humans, nevertheless humans cannot maintain it for any length of time without some adjustment of position. Walking is the easiest and most commonly used method of maintaining the upright position; when not walking (or running), humans lean, sit or lie down. The fact that ... the erect position is a tiring one for man, instead of being a state of rest, would disclose further in him an origin analogous to that of the other

mammals, if his organisation alone were taken
into consideration … The individuals of the
dominant race in question … had to multiply
their ideas … and thus felt the need for
communicating them to their fellows …
Individuals of the dominant race … stood in
need of making many signs, in order rapidly to
communicate their ideas … achieved the
formation of articulate sounds … habitual
exercise of their throat, tongue and lips in
the articulation of sounds will have highly
developed that faculty in them …

Such are the reflections which might be aroused
if man were distinguished from animals only by
his organisation, and if his origin were not
different from theirs.

This sentence was originally intended to be the closing sentence
of Part I of Lamarck's thesis. The arrival in Paris of a seal at the
end of June, 1809, just a few weeks before his book went to
press, caused Lamarck to make an addition to the subject matter
of Chapters VII and VIII.

After writing on seals for several paragraphs, Lamarck diverged to
speak of a number of other animals, from walruses to flying
squirrels. He concluded that water had been the cradle of the
entire animal kingdom. Three classes, the infusorians, polyps and
radiarians, still lived only in water and worms lived either in water
or very moist places.

… the worms appear to form one initial branch
of the animal scale, and it is clear that the
infusorians form the other branch … such worms
as are completely aquatic … become greatly
diversified in the water … those which
afterwards become accustomed to exposure to the
air have probably produced the amphibian

insects such as gnats, mayflies, etc., etc.
while these in turn gave rise to all the
insects which live altogether in the air.

Many insects, such as mosquitoes, live the first part of their lives in water before making the transition to an aerial existence, and many, even those who live their entire lives out of water, commence as some form of 'worm', a grub, a caterpillar or larva. Lamarck also suggested that the arachnids (spiders) originated in the water. The arachnids as we know them today live on land, but he suggested that, at an earlier stage in their evolution, they were related to the creatures which later became crustaceans. Reptiles had given rise to two main branches, one the birds, the second amphibian and, later, land mammals.

Lamarck offered a Table, showing that the animal scale began with at least two separate branches, and that as it proceeded it appeared to terminate in several twigs in certain places.

This series of animals begins with two
branches, where the most imperfect animals are
found; the first animals therefore of each of
these branches derive existence only through
direct or spontaneous generation.

These words should have alerted Elliot to the fact that he had been mistaken in his belief that Lamarck held that all existing animals had in their ancestry every existing species of animals. Lamarck clearly postulates two separate beginnings and several terminations.

In the Table, infusorians, polyps and radiarians are shown at the top right-hand side, as separate entities having no connection with other animals. Lamarck's complete text makes it clear that he believed all life had started as single-celled beings - *infusoria*.. Perhaps *infusoria* should have been a heading and *worms, polyps,* and *radiarions* shown as three separate groups? Be that as it may, Lamarck's Table illustrated the evoultion of only one group, *worms*, an evolution which resulted in many brances.

TABLE

SHOWING THE ORIGIN OF THE VARIOUS ANIMALS.

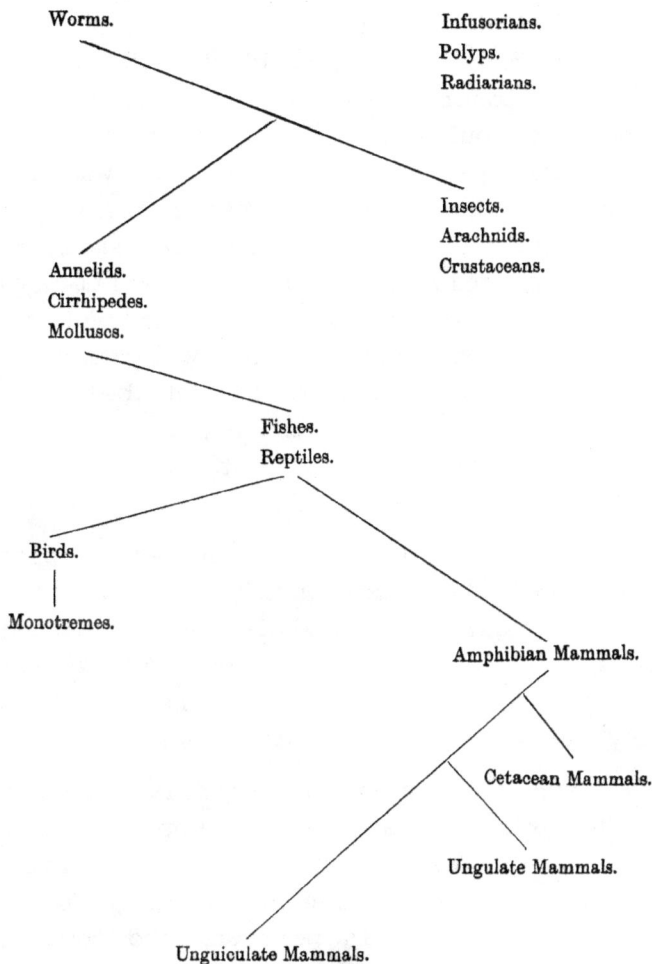

Lamarck concluded his additional notes with some thoughts about time:

If the duration of human life only extended to one second, and if one of our ordinary clocks were wound up and set going, any individual of our species who looked at the hour hand of this clock would detect in it no movement in the course of his life although the hand is not really stationary. The observations of thirty generations would furnish no clear evidence of a displacement of the hand for it would only have moved through the distance traversed in half a minute and this would be too small to be clearly perceived; and if still older observations showed that the hand had really changed its position, those who heard this proposition enunciated would not believe it, but would imagine some mistake, since they had always seen the hand at the same point of the dial …

Nature … only immutable so long as it pleases her Sublime Author to continue her existence – should be regarded as a whole made up of parts, with a purpose that is known to the Author alone, but at any rate not for the sole benefit of any single part.

Since each part must necessarily change and cease to exist to make way for the formation of another, each part has an interest which is contrary to that of the whole … In reality, however, this whole is perfect, and completely fulfills the purpose for which it is destined.

Section Three

PART II
Physical Causes of Life

Introduction

LAMARCK commenced Part II of his work with another Introduction. Some, he claimed, spoke of 'Nature' as if it were a special entity but nature was nothing but the sum total of all physical bodies and the laws which governed them. All physical bodies ... are liable to contract with one another various kinds of unions ... these bodies thus derive new properties and faculties from the condition in which each of them is placed ... There continually reigns throughout the whole of nature a mighty activity a succession of movements and transformation of all kinds ...

The idea of nature as eternal, and hence as having existed for all time, is for me an abstract opinion without foundation, finality or probability ... I imagine and like to believe in a First Cause or, in short, in a Supreme

Power which brought nature into existence and made it such as it is …

It is no doubt a very important matter to enquire into the nature of what is called life in a body; what are the conditions of organisation necessary for its existence … these are beyond question the most difficult to solve …

It seems to me that it is much easier to determine the course of the stars observed in space … than to solve the problem of the origin of life in the bodies possessing it … the difficulties are not insuperable; for in all this we have to deal only with physical phenomena … the mutual relations of different bodies … movements set up in the parts of those bodies by a force whose origin it is possible to ascertain.

Lamarck cited the work of a M. Cabanis who established great truth when he claimed that the moral and the physical both sprang from a common origin. By 'moral' was meant the non-physical, thought, intelligence, emotions, creativity … the operations called moral are directly due, like those called physical, to the activity either of certain special organs, or of the living system as a whole … it is in the simplest of all organisations that we should open our enquiry as to of what life actually consists, what are the conditions necessary for its existence, and from what source it derives the special force which stimulates the movements called vital … it is only by a study of the simplest organisations that we can attain a knowledge of the true conditions for the existence of life in a body.

Perhaps by now the reader will be coming to appreciate the reason why I questioned Elliot's claim that Lamarck had accepted the Chair of Invertebrate Studies because there was no other option open to him. It is possible that his deep interest followed the commencement of his work, but it is equally possible that he chose this work because he had already realized its importance and significance. Lamarck continued: we must enquire ... from what source living bodies derive the peculiar force which animates them ... it can no longer be doubted that this cause which animates living bodies is to be found in the environment of those bodies, and thus varies in intensity according to places, seasons and climates. It is in no way dependent on the bodies which it animates, it exists before they do and remains after they have been destroyed.

According to the account of creation given in the Bible in the book of Genesis, God created Man from the dust of the Earth and then breathed life into him. Lamarck said 'No'. Life was present everywhere first and manifested in bodies when they came together in a form which was able to support animation. He would shortly argue that this form was gelatinous or mucilaginous matter. In this part of his book, he is going to investigate how spontaneous generation can have occurred ... which are the living bodies which nature may produce spontaneously. Lamarck here suggested the argument he would later put more fully, that if spontaneous generation was possible at one point in time, then it must be possible at others.

It is to the influence of the movements of various fluids in the more or less solid substances of our earth that we must attribute the formation, temporary preservation, and reproduction of all living bodies observed on its surface, and of all the transformations incessantly undergone by the remains of these bodies.

The term 'fluid' was applied to any substance capable of movement. Besides water, air and other gases, it was known that heat could travel through solids. Heat moved through the iron bar in a forge. Heat was therefore fluid. Sound moved through air and, to a lesser extent, through some solids. It, too, was fluid. It was now being suggested that light, too, travelled, that it, too, was fluid. Gravity was a force unseen, unobserved and unfelt – until one was in the act of falling! Electricity was another, more recently discovered, force, which moved and was, therefore, fluid. Water and gases were known as visible fluids, the others as subtle fluids. Lamarck believed sunlight to be the principle substance incessantly causing modifications and displacements of great masses of these fluids in certain regions of the earth, and forcing them to undergo a kind of circulation and various sorts of movement, so that they are able to produce all the observed phenomena.

It was the movement of these fluids which cut out paths and establish[ed] depots and exits, to create canals and afterwards various organs in the gelatinous and mucilaginous material from which animals and plants were formed.

Living bodies thus constitute, by their possession of life, nature's principal means for bringing into existence a number of different compounds which would never otherwise have arisen.

It is vain to imagine that living bodies find readily formed in the substances on which they feed all the material required for building up the various parts of their bodies; they only find in these food substances, materials suitable for entering into the combinations which I have mentioned, and not the combinations themselves …

Such are the subjects which compose the second part of this work …

Chapter I – Inorganic v. living bodies

ALTHOUGH three kingdoms, animal, vegetable and mineral, had long been recognized, there was, at the time Lamarck was writing, a body of opinion which claimed that there the were entities which blurred or straddled these boundaries, having characteristics of more than one type. Lamarck disagreed with this opinion and devoted the first chapter of Part II of his work to an analysis of the three kingdoms, drawing attention to their differences, similarities and, in the case of plants and animals, to their analogous evolution.

Lamarck made no claim to be a chemist, although in the previous chapter he had described himself as a naturalist and a physicist, but his interest in plants had clearly led him to ponder the relationship between the organic and inorganic worlds: I long ago conceived the idea of making a comparison between organized, living bodies and crude, inorganic bodies … if we wish to arrive at a real knowledge of what constitutes life… we must above all pay very close attention to the differences existing between inorganic and living bodies.

> (1) No crude or inorganic body possesses individuality … Every living body, on the other hand, possess an individuality throughout its mass and volume.
>
> (2) An inorganic body may present a truly homogenous mass or it may constitute a heterogeneous mass … All living bodies, on the contrary, even those with the simplest organisation, are necessarily heterogeneous.

(3) An inorganic body may constitute either a perfectly dry, solid mass or a completely liquid mass or a gaseous fluid ... no body can possess life unless it is formed from two kinds of necessarily co-existing parts, the one solid, but supple and capable of holding liquids; the other liquid and contained in the first, but quite independent of the invisible fluids which penetrate the body and develop within it. Inorganic bodies have no specific special shape ... their shape does not remain permanently the same ... Living bodies, on the contrary, nearly all exhibit a shape peculiar to their species and one which cannot vary without giving rise to a new race.

(4) The integral molecules of an inorganic body are entirely independent of one another ... the molecules of a living body ... are dependent for their character upon one another

(5) No inorganic body needs any movement of its parts for its preservation ... so long as the parts remain at rest the body is preserved without disintegration and might exist in this condition for ever ... Every body possessing life ... is permanently or temporarily animated by a special force, which incessantly stimulates movements in its internal parts ... effecting restorations, renewals, developments and a number of phenomena that are entirely peculiar to living bodies

(6) In all inorganic bodies an increase of volume and mass is always accidental and has no necessary limits ... The growth of every living body ... is always necessary and limited, and only takes place by intussusception, that is to say, by internal penetration

(7) No inorganic body has to feed in order to be preserved ... No such [living] body can maintain life if it is not constantly feeding ...

(8) Inorganic bodies ... are not born ... All living bodies ... are really born, and are the produce either of a germ which has been vivified and prepared for life by fertilisation, or else simply of an expandable bud ... new individuals arise exactly like those which have produced them

(9) No inorganic body can die ... Death is a necessary result of the existence of life.

How great is the error of those who try to find a connection or sort of gradation between certain living bodies and inorganic bodies!

Lamarck then turned his attention to the comparison of plants and animals, commencing with their common characteristics. The only point in common between animals and plants is the possession of life ... in both cases they are bodies composed essentially of two kinds of parts, the one solid but supple and containing; the other liquid and contained, but independent of the invisible fluids which penetrate and develop within them ... possess individuality ... have a shape peculiar to their species ... are permanently or temporarily animated by a special force which stimulates their vital movements; are only preserved through nutrition ... grow for a limited period by internal development ... reproduce and multiply ... reach a period when the state of their organisation no longer permits of the maintenance of life within them.

The reader may be puzzled by Lamarck's statement that organic bodies may be 'permanently or temporarily animated'. This point will become clear later.

Lamarck proceeds to a comparison of the characteristics distinguishing plants from animals.

Plants are organised living bodies, not irritable in any of their parts, incapable of performing sudden movements several times in succession … In animals, some or all of the parts are essentially irritable, and have the faculty of performing sudden movements which may be repeated several times in succession … The only sudden movements that certain plants display are movements of relaxation or collapse … other movements performed by the parts of plants, such as those which make them bend towards the light, which cause the opening and closing of flowers … are carried out so slowly as to be altogether imperceptible …

Plants … grow … in two opposite directions … ascending … and … descending vegetation … start from a common point … named the vital knot … the plant only really dies when life ceases to exist in this part … in animals … their development is not limited to two special directions only, but takes place on all sides and in all directions, according to their requirements … life is never concentrated in an isolated point …

The food of plants consists only of the liquid or fluid substances which they absorb from the environment … they never have to carry on digestion … living bodies themselves elaborate their own substances … they … form the first non-fluid combination … Most animals … feed on substances which are already compound … Hence they have a digestion in order to bring about the complete solution of the substances … they … form the most complex combinations …

In plants … solids exist in larger proportion than fluids … in animals fluids are more abundant than solids …

Lamarck concluded this chapter by drawing attention to some analogies in the evolution of plants and animals.

In both of them, the most simply organised only reproduce by gemmæ or buds … which … require no preliminary fertilisation … in both plants and animals, when the complexity of organisation was sufficiently advanced to permit of the formation of organs of fertilisation, the reproduction of individuals then became exclusively or chiefly sexual. [Quite why, or how, this occurred is still not understood.]

In the winter of cold climates … the woody perennial plants undergo a more or less complete suspension of vegetation … during these conditions, there occur in the plants no losses or absorptions of food, or any alterations or development … In the winter of cold climates … great many [animals] become more or less completely torpid … just as there are simple animals … and compound animals adhering together … sharing a common life such as most of the polyps, so also there are simple plants living as individuals and there are compound plants where several live together, are grafted on to one another and share a common126 life …

If the name of zoophyte were given a century ago to compound animals of the class of polyps, the error was excusable; now things have altered … Let us now enquire what life is, and what are the conditions for its existence in a body.

Chapter II – Of What Life Consists

LAMARCK commenced this chapter with a recapitulation of his conclusion that life could only manifest when there was an interactive relationship between three components: supple, containing parts, a fluid contained within those solid parts capable of movement and an 'exciting cause' of such movement. It is … from the relations between these three objects that the movements, changes, and all the phenomena of life result. There was a necessity for an exciting cause and for the body to be capable of responding to that exciting cause.

Lamarck then explained what he had meant in his previous chapter about life being present 'permanently or temporarily'.

If fluid matter were artificially removed from simple organisms, such as polyps or infusoria, they became desiccated, to all intents and purposes 'dead'. However, if the internal fluids were returned to these bodies, and a gentle warmth applied, movement, life, was soon restored, showing that life had been but temporarily suspended. The same occurred with simple plants, such as mosses. A return to life after apparent death was possible even in humans, although only when the person had been 'dead' for a short time. There had been occasions when people had been pulled from the water after having been submersed for up to an hour, apparently dead, but who had been restored to life if stimulus by means of contractions had been applied which resulted in movement of the internal organs, causing circulation to be restored.

Considering only the more complex living bodes would not make clear the conditions necessary for life. These could only be established by the study of the least complex animals and plants.

Considering only animals and plants at the most simple extremity, it became clear that in each individual the body consists only of a gelatinous or mucilaginous mass of cellular tissues of the feeblest coherence, the cells of which are in communication, and the various fluids of which undergo movements, displacements, dissipations, subsequent renewals, changes of state, and finally deposit parts which become fixed there … an exciting cause of varying activity, but never entirely absent, incessantly animates the very supple containing parts of these bodies, as well as the essential fluids contained in them …

In animals, the exciting cause of organic movements acts powerfully both on the containing parts and on the contained fluids … In plants, on the contrary, the exciting cause in question only acts powerfully on the contained fluids … but its only effect on the containing parts of these living bodies … is an orgasm or slight erethism which is too feeble to permit of any movement or to cause a reaction on the contained fluid or consequently to endow these parts with irritability

… as the vital energy increases in proportion to complexity of organisation, there soon arrives a time when irritability and the exciting cause are no longer sufficient by themselves for the acceleration needed in the movement of the fluids; nature then makes use of the nervous system …when this system permits of muscular movement, the heart becomes a powerful motor for accelerating the movement of the fluids …

Lamarck then reinforced the point he had made earlier – that just because a certain set of conditions were necessary for the maintenance of life in one organization did not mean that those same conditions were necessary for life in all organizations. Complex organisms require many organs to sustain life; simple organisms require but few; some require none.

Lamarck then moved to another, related, topic. He pointed out that there were two different types of 'nervous influence', that which128 produced movement and that which produced feeling. The nervous influence which produced movement in muscular tissue did not produce feeling in them. Under normal circumstances, we are not aware of the movement of the muscles of our hearts. If our hearts are beating unusually strongly for some reason, we will become aware as the heart muscle strikes neighbouring sensitive parts ... when we walk or perform any action we never feel the movement of the muscles nor the impulses which drive them.

Based on this, Lamarck rejected the axiom 'Living is feeling'. Plants, to the best of Lamarck's knowledge, lived but they did not feel.

Chapter III– The Exciting Cause of Organic Movement

A T THE time that Lamarck was writing, electricity was a (comparatively) new phenomenon. Very soon after its discovery, an association had been suggested between this new source of energy and the energy manifested in the nervous system. Lamarck accepted this association and in this chapter he explained his reasons for believing that caloric and electricity were the two principle fluids necessary for life.

Caloric meant more than heat as measured by a thermometer. It was the term used at that time for the 'fluid' which manifested as heat. Warmth or heat was the result of the presence of caloric. For the rest, this chapter is mainly self-explanatory and needs little by way of comment.

The ancient philosophers felt the necessity for a special exciting cause of organic movements … they imagined a vital principle, a perishable soul for animals, and even attributed the same to plants … they created mere words to which are attached only vague and unreal ideas

… The only knowledge that it is possible for us to acquire is and always will be confined to what we have derived from a continued study of nature's laws; beyond nature all is bewilderment and delusion: such is my belief

… I do not wish to go back to the consideration of first causes … We shall confine ourselves to a study of the immediate recognised causes acting on living bodies

… It would doubtless be impossible to ascertain the exciting cause of organic movement if the subtle, invisible, uncontainable, incessantly moving fluids which constitute it were not disclosed to us in a great variety of circumstances; if we had not proofs that the whole environment in which all living bodies dwell are [is] permanently filled with them … if we did not know positively that these invisible fluids penetrate more or less easily the masses of all of these bodies and stay in them for a longer or shorter time … some of them are in a constant state of agitation and expansion, from which they derive the faculty of distending the parts in which they are insinuated.

It will be remembered that in the previous chapter Lamarck spoke not only of the visible fluids, such as water and gases, but of the subtle fluids, such as warmth, magnetism, as well as of the (again comparatively) new field of energy, known as gravity. Subtle fluids penetrate all solid bodies, with varying degrees of ease. Everything, both organic and inorganic, is subject to gravity. Lamarck is going to argue that all organic bodies are penetrable by electricity in much the same way as they are by gravity.

… it is well known that … no part of the earth inhabited by living beings is destitute of caloric (even in the coldest regions), of electricity, of magnetic fluid, etc. …

We do not yet know how numerous may be these subtle invisible fluids which are distributed in constant agitation throughout the environment … these invisible fluids penetrate every organised body … They thus stimulate movements and life, when they come in contact with an order of things permitting of such results … two of them appear to us to be the essential elements of this cause, viz. caloric and electric fluid. These are the direct agents which produce orgasm and the internal movements which in organised bodies constitute and maintain life.

Caloric appears to be that of the two exciting fluids in question which causes and maintains the orgasm of the supple parts of living bodies; the electric field is apparently that which provides the cause of the organic movements and activities of animals …

It is possible, however, that other active invisible fluids combine with the two already named in the composition of the exciting cause.

… In animals of low organisations, the caloric of the environment seems to be sufficient by itself for the orgasm and irritability of their bodies … The case is different with regard to animals of highly complex organisations: the caloric of the environment merely completes or rather aids and favours the power which these living bodies themselves have of constantly producing caloric within them [i.e. warm blooded animals].

… It appears … that this electric fluid, which is introduced through the medium of respiration or of food, has undergone some modification in the animal's interior and become transformed into nervous or galvanic fluids.

… a great reduction of temperature would exterminate all living bodies long before reaching the point of absolute cold … we know that no part of the earth's surface and at no period of the year do we ever find a total absence of caloric.

… without a special exciting cause of orgasm and vital movements … life could not exist in any body. Now this exciting cause has nothing to do with the visible fluids of living bodies, nor with the solid containing parts of these bodies.

… This same exciting cause is also the cause of fermentation, the manifestations of which it alone brings about in all compound non-living matter, whose parts are favourable to it. Thus in great reductions of temperature the activities of life and fermentation are suspended more or less completely, in proportion to the intensity of the cold.

… Although life and fermentation are two very different phenomena, they both derive from the same origin the movements by which they are constituted … in bodies possessing life, the existing order and state of things are such that every decomposition of principles is subsequently made good by new and closely similar combinations as a result of continued movements, whereas in the unorganized or disorganised fermenting body, the decompositions which occur cannot be made good by a continuance of fermentation.

… As soon as the individual dies, its body, which is then disorganized … immediately joins the class of bodies liable to fermentation … The exciting cause which gave it life then hastens the decomposition of such of its parts as are capable of fermentation.

… exciting causes of vital movements must necessarily be sought in the invisible, subtle,

penetrating and ever active fluids with which
the environment is always supplied …

Lamarck then writes of how an increase in temperature, up to a certain point, encourages greater growth and an increased rate of reproduction. Life is far more prolific in tropical climes than in polar. Too much heat is inimical to life, especially in the absence of water. Water is more plentiful in tropical areas than it is in temperate and this aids in the proliferation of life. This is true of both plants and animals. Plants also need light, although some animals are able to exist in dark environments. Were it not for animals eating plants and other animals, the earth would be completely smothered in living organisms.

Lamarck concludes this chapter by re-iterating that it may be that some other invisible fluid combines with electricity in making up the cause which is able to excite vital movements … but I see no reason for supposing it. It seems to me that caloric and the electric substance together are quite sufficient to constitute the essential cause of life … This apparently is the mode of action of the exciting cause of life; but it cannot be regarded as established, until it is possible to find proofs of it.

Chapter IV – Of Orgasm and Irritability

LAMARCK commenced this chapter by alerting his reader to the fact that the 'orgasm' of which he would here be speaking was not that of a sexual nature but rather that which was sometimes referred to as 'erethism'. This was a general condition, characteristic of the supple, internal parts of animals and essential for life.

This chapter does not greatly advance Lamarck's main theme, his argument that all life on this Earth has evolved. Rather it is of interest because it illustrates the lack of knowledge still hampering physiologists with regard to the characteristic of warm blood. Lamarck had previously pointed out that at no point on this planet was there a complete lack of caloric, nowhere was there absolute cold, or absolute zero to use the terminology of today. The same was true of plant and animal life. 'Warm blooded' was a relative term, since no physical fluid flowing through the body of any plant or animal was ever at freezing point, let alone below it. Plants and cold blooded animals are more dependent on an external source of heat than are warm blooded. Plants cannot survive in polar regions, nor can cold blooded animals, but some warm blooded ones can. Cold blooded animals need to bask in the sun to increase their metabolism but, even then, they are not able to sustain the same level of activity as warm blooded animals.

Some warm blooded animals hibernate during cold weather. When the temperature rises, they return to activity. However, if the temperature drops further than comfortable for the animal in this state, it does not drift into death. First, the increased cold causes an irritation in their nerves which wakens and agitates them, and revives their organic movements and hence their internal heat. Presumably this phenomena is some form of shivering. If the temperature remained too cold, the animal would die. Lamarck was not sure whether this was the case with cold blooded animals – he thought it probably was – but he knew the case was different with simple animals without nerves. Without sufficient external warmth, their activity slowed and continued to slow until death.

Lamarck devoted this chapter to trying to explain these phenomena, from where came internal warmth, how did its presence affect life, what was its action?

Lamarck held that the 'tone' of the supple parts of the bodies of both plants and animals was maintained by an invisible, expansive, penetrating fluid (possibly several) ... with a strength that is proportional to the favourable disposition of the parts; it is stronger according as they are more supple and less dried up. When plants became dry, they died. This may be the result of lack of water, but no plant lives forever and when the invisible fluid was insufficient, or absent, no amount of watering would revive the plant. Presence of a physical fluid alone was insufficient to maintain the orgasm, erethism, tone, or whatever term one chose to use ... the death of an individual gives rise to a relaxation and subsidence of the supple parts, making them softer and more limp than in the living state ... an expansive caloric is continually emanating from the arterial blood of many animals and constitutes the principal cause of orgasm in their supple parts.

Caloric is present everywhere in some degree. Lamarck believed caloric to be the principle subtle fluid maintaining orgasm. It would seem that chemists had not yet established that every chemical reaction produces heat, because Lamarck made no suggestion that bodily functions, such as digestion, could be contributing to the heat in an animal's body. He did, however, accept that there may be a connection between respiration and warmth, but not as strong a connection as was being suggested by some of his colleagues.

He accepted that during respiration a gas was absorbed into the blood and further accepted that this gas was probably oxygen. Lamarck accepted that when the bodily temperature was elevated in a sick person, that person was often observed to breathe more rapidly but rejected the idea that it was this increased rate of respiration which caused the elevation of temperature. Local inflammation, such as a boil, was associated with an increase in temperature in that part clearly not associated with any increase in the rate of respiration. This led Lamarck to argue that a component of the blood was 'freed' by combination with the inhaled oxygen and that its liberation produces our internal heat.

The more complex the animal, the more it was possessed of supple, internal organs, the more subtle fluid it was capable of receiving. This (or these) subtle fluid(s) interacted with the physical fluids and in the more complex animals resulted in an increase in bodily temperature.

An expansive subtle fluid was everywhere present. It invested all living matter and was constantly seeking new pathways to manifest. This, Lamarck had previously explained, caused channels to be formed, giving rise to the nervous and circulatory systems. This expansive energy was also responsible for the formation of all internal organs. Its increased presence in the more complex animals resulted in an increase in bodily temperature. Lamarck was convinced that the subtle fluid, known as caloric, was primarily responsible for the evolution of living forms. The similarity in the thinking of Lamarck with that a century-and-a-half later of Teilhard de Chardin has already been noted.

'Irritability' in animals resulted in a sudden contraction and shrinkage of the irritated point ... followed by a contrary movement … so that the natural condition of the part distended by orgasm is promptly reestablished. This contraction may be the result of external touch (involuntary) or of internal action (voluntary) as the result of nervous impulse. Muscles operate by temporary contraction but must be returned to their relaxed condition as soon as possible. It is possible to extend the period of contraction by force of will for a period of time but not indefinitely. This phenomenon is obviously quite different from that of sensations. Some measure of orgasm (erethsm) was present in all living forms, both plant and animal, whereas sensation was only present in animals with a nervous system. Lamarck postulated that some form of fluid (gas) was released from tissues which caused them to contract but which was rapidly replaced by fluid (gas) released from the blood. Orgasm was thus a combination of a subtle fluid and a physical fluid.

Irritability ceased after death. In humans, this occurred within two or three hours of death; the heart of a frog was still capable of being irritated and producing movement for up to thirty hours after death; in insects, irritability may last even longer post mortem.

Chapter V – Of Cellular Tissue

N THIS chapter, Lamarck further expanded upon material he first put forward in an address in 1806. While it was no new had discovery that all the organs of animals are invested by cellular tissue, even down to their smallest parts … yet … no one that I know of has yet perceived that cellular tissue is the universal matrix of all organisation … In 1806, Lamarck had claimed that the movement of fluids through it [cellular tissue] is nature's method of gradually creating and developing … organs out of this tissue.

It will be remembered that at this time microscopes were not yet sufficiently advanced to distinguish sub-cellular material; cells were still thought of as characterless pieces of gelatinous or mucilaginous matter, both small and insignificant. Even though Lamarck had no concept of the complexity of cellular material, he had intuitively grasped that cells were the fundamental building blocks of life.

… in the natural order, both of animals and plants, those living bodies whose organisation is the simplest and which are consequently placed at the extremity of the order, consist only of a mass of cellular tissue in which there are to be seen neither vessels nor glands nor any viscera …

Take for instance the algæ … their scarcely modified cellular tissue is conspicuous enough to prove that it alone forms the whole substance of these plants. In several of these algæ … internal fluids … have not given rise to any signs of a special organ... others ... have only traced out a few canals through which food is supplied to those reproductive corpuscles, which botanists take for seeds

... in the most imperfect animals, such as the infusorians and polyps, and in the least perfect plants, such as the algæ and fungi, there sometimes exists no trace of any vessels and sometimes only a few rudimentary canals ... the very simple organisation of these living bodies consists only of a cellular tissue, in which slowly move the fluids which animate them ... But everywhere, even in the most perfect animals, there is really nothing in their interior but a cellular tissue modified into a large number of different tubes.

Lamarck then reminded his reader how much more simple were the tubes in plants than in complex animals. It was the movement of fluids within the cellular tissue which drove evolution. Lamarck cited a passage from his Recherches sur les corps vivant:

... every organisation and every new shape, acquired by this agency and contributing circumstances, were preserved and transmitted by reproduction, until yet further modifications had been acquired by the same method and in new circumstances.

This sentence encapsulates Lamarck's theory. Lamarck knew nothing of genetics. Not for him a fortunate mutation at the time of conception which produced, not only a viable embryo, but an advantaged one which, happily surviving the vicissitudes of infancy, grew into a adult whose progeny would multiply and grow into a new species. For Lamarck, change was orderly, ever following the same natural laws and pathways, initiated by further modifications ... acquired by the same method and in new circumstances. 'The same method' was the circulation of subtle fluids in the organisation. Physical fluids could only follow channels initiated by the subtle fluids. 'New circumstances' were a change in the environment. This could have been a change in temperature or a change in landscape, the latter including the organic flora and fauna inhabiting an area and thus forming part of the environment. Once established, life forms remained stable until faced with 'new circumstances' when modifications would take place in accordance with the same, previously operating, natural laws.

In some ways, Lamarck's theory pre-empted that of Niles Eldredge and Stephen J. Gould, who in 1972 published their theory of Punctuated Equilibrium. Eldredge and Gould claimed that, rather than evolution occurring gradually and steadily as claimed by Darwin, the fossil record showed that forms remained stable for vast periods of time and then were subject to sudden change. There was resistance to this view by those who thought Eldredge and Gould were suggesting saltation – sudden evolutionary change by 'leaps' or 'jumps', something denied by all Darwinists. Eldredge explained that what appeared sudden in the fossil record may have taken place over tens of thousands of years. So long as conditions remained the same, there was equilibrium and forms remained the same. Change in form followed a change in conditions. Eldredge, and others, established that extinctions followed times of cooling, sometimes moderate, sometimes severe, resulting in Ice Ages, the most severe of which resulted in mass extinctions. As temperatures gradually rose, new forms evolved to replace those which had disappeared. Lamarck would have been pleased to see his theory that 'caloric' (warmth) was a major factor in the evolution of new forms, supported!

Lamarck continued by claiming it followed that this movement of fluids has ... the faculty of gradually increasing the complexity of organs ... according as new modes of life or new habits ... create a necessity for new functions ...

This was no giraffe, looking longingly at a leaf just out of its reach, thinking to itself:

> By heck, by heck,
> I need a longer neck!

Living forms remained constant so long as conditions remained constant. Changes in the environment (including changes in flora and fauna which changed the environment, quite apart from global or local changes in temperature or geology) resulted in new conditions and new needs. All plants or animals in a population would be subject to the same changed conditions. The subtle fluids would act on all members of the population in the same way and changes would gradually take place. Quite how, or when, these changes became established to such a degree that they became subject to the normal processes of reproduction, Lamarck did not know. Lamarck merely accepted that at some point, this occurred.

Reference has already been made to Sheldrake's theory of morphic fields, supported by Hollick (2006), which proposed that energy fields, emitted by both organic and inorganic matter, were an important part of the environment which effected the development and evolution of all living substances. This theory echoed that of Lamarck, in that it claimed invisible force fields/subtle fluids were the driving force behind evolution, not chance genetic mutation.

Chapter VI – Spontaneous Generation

THIS chapter is one of the most important in the entire book since it deals with one of the most profound questions philosophers ask – how did life on Earth begin?

In his first sentence, Lamarck once again made reference to the Supreme Author of all things. He claimed that life and organization on Earth were natural phenomena which had come into being through the operation of powers conferred on nature by the Supreme Author and in accordance with the laws by which nature was constituted. He pointed out that ancient philosophers had observed the power of heat in initiating/sustaining life but had failed to note that the presence of moisture was also necessary.

Ancient philosophers had also recognized that direct (spontaneous) generations occurred, with which Lamarck agreed, but he disagreed with them about the level of organization at which this was possible. They had believed that all the animals then designated worms … are born in favourable times and places, as a result of the action of heat on various decaying substances. It was then believed that putrid flesh directly engendered larvæ, which were subsequently metamorphosed into flies … Science had now acknowledged that all animals, no matter how imperfect, had the power of reproduction. Furthermore, all insects and animals of higher organization reproduced by sexual generation, while polyps reproduced by means of gemmae or buds. These discoveries had led scientists to deny that spontaneous generation ever occurred.

This conclusion is erroneous in being too universal; for it excludes the direct generations wrought by nature at the beginning of the animal and vegetable scales ... Not only has it been impossible to demonstrate that the animals with the simplest organisations, such as the infusorians and among them especially Monas, and also the simplest plants such perhaps as the Byssus of the first family of algæ, have all sprung from individuals similar to themselves ... there are observations which go to show that these extremely small and transparent animals and plants, of gelatinous or mucilaginous substance, of very slight coherence, curiously ephemeral, and as easily destroyed by environmental changes as brought into existence, are unable to leave behind them any permanent security for new generations. ... nature ... in bringing various bodies into existence ... must necessarily have begun with the simplest ...

Lamarck claimed that all life commenced in a very simple, supple, form; excess nutrition enabled it to prepare material suitable for reproduction, a concept which seems strange today but which was generally accepted in past times. As life continued, bodies gradually hardened, again as the result of 'excess nutrition' which became deposited in the body, giving rise to its harder parts, at first a necessity for its development but eventually leading to a general hardening of the body, part of the process of old age, which eventually became so severe that life could no longer be sustained in that individual organism.

... life ... tends incessantly by its very nature towards higher organisation ... reproduction permanently preserves all that has been acquired ... from the remains left by each of these bodies after death, have sprung the various minerals known to us.

Elliot derided Lamarck's idea that minerals had come from the bodies of dead animals. I earlier stated that I was sure Lamarck had not meant to imply that vegetable and animal life had evolved on an Earth devoid of solid substances. So what exactly did Lamarck mean by 'mineral'? Clearly, he did not mean metals, such as gold and silver, which have no organic origin. However, he had drawn attention to the fact that, within organic bodies, nature brought together chemicals which formed new substances in a way which never occurred outside living organisms. These substances, on the death of the individual plant or animal, became part of the surface of the Earth. They may now be inert, but they are not inorganic. In former times, it had been assumed that the surface of the Earth now was much as it had been at the time of Creation a few thousand years ago, except for erosion, volcanoes, etc. Lamarck had pointed out that vast expanses of the earth's surface, including chalk cliffs and downs hundreds of feet high, had been formed from the remains of previously living creatures. I believe that sense can only be made of Lamarck's remarks regarding 'minerals' if the word 'organic' or 'compound' is placed before it.

Lamarck then returned to the main topic of this chapter. Nature only establishes life in bodies that are at the time in a gelatinous or mucilaginous state … every germ at the moment of its fertilisation … is in a gelatinous or mucilaginous state.

Lamarck then compared the two processes by which nature engenders life: fertilization and direct generation.

… in the reproduction of mammals, the vital movement in the embryo appears to follow immediately upon fertilisation; whereas in oviparous animals there is an interval between the act of fertilizing the embryo and the first vital movement induced by incubation … during this interval the fertilized embryo cannot yet be reckoned among living bodies; it is ready no doubt for the reception of life … but so long as organic movement has not been originated by this stimulus, the fertilized embryo is only a body prepared for the possession of life and not actually possessing it … In the same way, the seed of a plant … does not enclose a living embryo unless it has been exposed to germination. Germination occurs when the seed receives moisture and an appropriate degree of warmth. Seeds can remain dormant for years before germinating under the necessary conditions. Lamarck is claiming that life does not commence at the time the flower receives fertilization by means of a visiting insect, or by the power of the wind. Life only begins at germination which takes place in the presence of warmth and moisture. If there is no incubation of an egg or germination of a seed, there will be disintegration, but no death, since there has been no life. Fertilization is not the commencement of life, but a preparation for life.

… it is enough that a subtle penetrating vapour, which escapes from the fertilizing material should be insinuated into the gelatinous corpuscles capable of receiving it; that it should spread throughout its parts and by its expansive movement break up the adhesion between these parts and so complete the organisation already begun and dispose the corpuscles for the reception of life, that is of the movements constituting life. Lamarck called this penetrating vapour aura vitalis.

It may be remembered that to this day science does not know how or why the coming together of certain strands of DNA from two separate individuals results in the birth of another living being. Some babies are still born. There may never have been a heartbeat but growth continued, sometimes to full term, on the 'borrowed' vitality of the mother. Fertilization, alone, does not guarantee a live birth. It is, as Lamarck claimed, but a necessary step in the process. In reproductive fertilization, the 'fertilizing vapour' breaks up the rudiments of the organization which no longer should adhere together and re-arranges them in a particular way. In direct (spontaneous) generation, the subtle surrounding fluids are introduced into the mucilaginous or gelatinous material, transforming it into a mass of cellular tissue by its expansive action. If the material is gelatinous, animal life will follow; if it is mucilaginous, then only vegetable life will be able to exist in it.

We cannot penetrate further into the wonderful mystery of fertilisation ... we must first remember that a subtle penetrating fluid in a more or less expansive condition, and apparently very analogous to the fluid of the fertilizing vapour, is distributed everywhere throughout the earth

... *Nature, by means of heat, light, electricity and moisture, forms direct or spontaneous generations in that extremity of each kingdom of living bodies, where the simplest of these bodies are found.*

... light is known to generate heat, and heat has been justly regarded as the mother of all generations. These two distribute over our earth at least, the principle of organisation and feeling; and since feeling in its turn gives rise to thought as a result of the numerous impressions made on its organ by external and internal objects through the medium of the senses, the origin of every animal faculty may be traced to these foundations.

Life has now been found at the deepest, darkest depths of the ocean – near thermal vents! This would seem to indicate the warmth, Lamarck's first preference, is more important than light in the establishment of life. Indeed, many seeds, be they plant or animal (within the egg or uterus) develop in the absence of light.

If we consider the most imperfect of … animals, such as the infusorians, we shall see that in a hard season they all perish, or at least those of the most primitive orders … from what or in what way do they regenerate? … Must we not think that these simplest organisms, these rudiments of animality, so delicate and fragile, have been newly and directly fashioned by nature rather than have regenerated themselves? … suitable portions of inorganic matter, occurring admist favourable surroundings, may be the influence of nature's agents, of which heat and moisture are the chief, receive an arrangement of their parts that foreshadows cellular organisation and thereafter pass to the simplest organic state and manifest the earliest movements of life …

Nature, by means of heat, light, electricity and moisture, forms direct or spontaneous generations at that extremity of each kingdom of living bodies, where the simplest of these bodies are found.

The body that is most fitted for the reception of the first outlines of life and organisation is any mass of matter apparently homogenous, of gelatinous or mucilaginous consistency, and whose parts though cohering together are in a state closely resembling that of fluids, and have only enough firmness to constitute the containing parts.

Since Lamarck wrote these words, many scientists have endeavoured to generate life in the laboratory, using methods simulating the conditions suggested by Lamarck. None have been successful and so the mystery continues.

> ... it appears to me certain that nature does herself carry out spontaneous or direct generations, that she has this power, and that she utilizes it at the anterior extremity of each organic kingdom, where the most imperfect living bodes are found; and that it is exclusively through their medium that she has given existence to all the rest.

Charles Darwin, along with his peers, rejected spontaneous generation. Darwin also rejected Special Creation, as suggested by *Genesis*, thereby leaving his theory in somewhat of a quandary. At the beginning of his book, *On the Origin of Species*, Darwin stated that his theory only encompassed the origin of species, or varieties, which he considered to be the same. By the end of his book, it was clear to Darwin that he had argued himself into a position where the logical extension of his theory was that all life had evolved from one, or a few, very simple organisms, but he declined to speculate on how this life had come into existence.

Lamarck was brave enough, and thorough enough, to tackle this very difficult philosophical/scientific question. He was writing before Pasteur and, while he may not have been right, no one else has yet solved this mystery. It is up to science to demonstrate, not only that he was wrong, but how he was wrong.

At the beginning of the chapter, Lamarck had stated that earlier scientists had been wrong when they claimed that worms spontaneously generated, for example on rotting meat. At the end of the chapter, Lamarck offered a caution. He admitted that his theory was not yet proven. While he was of the opinion that spontaneous generation occurred only at the very beginning of the animal scale, there was a possibility that he was wrong. Many people believed that, at the very least, intestinal worms must be created spontaneously in the gut. How else could they get there? Lamarck had to admit that he must keep an open mind in this regard, because the truth had not been proven either way.

In his early descriptions of his classes of animals, Lamarck had been uncertain regarding the reproduction of worms. The three groups, radiarians, polyps and infusorians all self-propagated by simple division or budding. Insects, he insisted, were all oviparous. Insects have larvæ, caterpillars, maggots, etc., which resemble worms, but they are unable to reproduce themselves. Only when they have metamorphosed, is it possible for them to reproduce. Worms, he believed, self-propagated by gemma, which did not need external fertilization, but these gemma seemed to be located inside the worm's body. Quite how new worms were formed and 'born', Lamarck did not know. They seemed to fill some intermediate place but quite what that place was, Lamarck was unable to determine.

Chapter VII – Results of Life

FOR once, I am in agreement with Elliot. This chapter is repetitive! It is as if Lamarck felt so strongly about the point he was making that he needed to repeat it several times, in a slightly different way, to make sure that his reader had fully grasped it.

Lamarck started by recapitulating the point he made in the previous chapter, and earlier, that living bodies were chemical factories. Within them, nature assembled chemical compounds which could never come together in any other way. He then went on to say that, in the past, savants had believed that there were two different sets of laws operating in nature, one for living matter, the other for nonliving. No! said Lamarck. Nature has but one set of laws which operate universally. Different results accrue because of the different characteristics of the two substances. Inorganic matter is hard; organic matter is supple.

One and the same cause necessarily has varied effects when it acts upon objects of different nature and in different conditions … There is no difference in the physical laws, by which all living bodies are controlled, but there is a great difference in the circumstances under which these laws act.

Lamarck had made it clear earlier in his book that he would not be discussing First Causes. At no point does Lamarck attempt to offer any explanation for the existence of the Universe in general or this Earth in particular. He accepts the (unexplained) existence of a Supreme Being, the Author of all things. Lamarck makes no attempt to explain the coming into existence of the first globules of gelatinous or mucilaginous material, in their warm and moist environments. His thesis concerned only the evolution of life after these things had occurred. Darwin later gave himself the same exemptions, claiming that First Causes were the province of the philosopher, subsequent events the province of the scientist.

The subtle, expansive fluids everywhere present could not act upon inorganic matter because it was hard. Organic matter was supple. The coming together of inorganic chemicals in such a way that they formed a supple compound, either gelatinous or mucilaginous, was the prime requirement for the establishment of living matter. Lamarck never speculated as to whether these two substances appeared at the same time, or, if not, which appeared first. However, towards the end of the chapter, Lamarck stressed the fact that plants were able to absorb the nutrition they needed directly from their environment, while animals survive on plants, or decayed plant products, so it is to be assumed that mucilaginous matter preceded gelatinous.

In living matter, the force of nature was synthetic; in non-living matter it was analytic. It tended to decompose or to degrade substances, breaking them down into their simple components. Both orders acted within the living body: there does exist in these bodies throughout life a perpetual struggle between those conditions which make the vital force synthetic, and those others, always being renewed, which make it analytic. If living bodies were only synthetic, if they only ever expanded, they would become unmanageable. Only by being counteracted by the analytic can the synthetic be sustained, although but temporarily, since eventually the analytic overpowers the synthetic, the living body 'grows old', deteriorates and dies.

Lamarck summed up the events which occurred thus: Change is constant in living bodies; real losses occur in their substance; there is a constant need to make good these losses through ingestion; these processes result in new combinations of materials which would never occur outside a living body.

Rather in the same way that the Chinese believed in the opposing forces of ying and yang, and the Indians believed in Shiva/Shakti, Lamarck believed there were two opposing forces operating in nature, that which destroyed products and that which built them up. Surprisingly, he listed them in that order.

After several more pages of somewhat repetitive writing, Lamarck finally assembled his arguments.

… the organic functions of all living beings confer on them the faculty, in some cases (plants) of forming direct combinations, that is of uniting free elements after modification, and of immediately producing compounds; in other cases (animals) of modifying these compounds and altering their character by the addition of new principles to a remarkable extent.

I must again stress the fact that living bodies form for themselves, by the activity of their organs, the substance of their bodies and the various secretions of their organs; and that they neither find this substance ready formed in nature, nor the secretions, which come purely from them alone.

It is by means of food, which animals and plants are obliged to use for the preservation of their life, that the organs of these living bodies work their effects. These effects consist in a modification of the food resulting in the formation of special substances, which would never have existed without this cause …

Thus plants, which have no intestinal canal nor any other organ for digestion, and which consequently use for food only fluid substances or substances whose molecules are not aggregated (such as water, atmospheric air, caloric, light, and the gases that they absorb), yet form out of such material, by means of their organic activity, all the juices that are proper to them, and all the substances of which their body is composed; that is, they form for themselves the mucilages, gums, resins, sugar, essential salts, fixed and volatile oils, feculæ, gluten, extractive and woody matter; all of them substances arising direct from immediate combinations, and none of which can ever be formed by art …

Animals cannot built up direct combinations like plants; hence they use compound substances for food; they have to carry out digestion (at least nearly all of them), and they consequently have organs for this purpose.

But they also form for themselves their own substance and secretions … out of grass or hay the horse forms by the action of its organs its blood and other humours, its flesh and its muscles, the substance of its cellular tissue, vessels and glands, its tendons, cartilages and bones, and lastly the horny matter of its hoofs, and the hair of its body, tail and main.

… vegetable productions are in general different from animal productions; and among the latter the productions of vertebrates are in general different from those of invertebrates …

I shall again repeat that if it is true, as can hardly be doubted, that all compound mineral substances such as earths and rocks, and all metallic, sulphurous, bituminous, saline substances, etc., arise from the remains of living bodies, - remains which have undergone successive decomposition on and under the surface of the earth and waters; it is equally true to say that living bodies are the original source from which all known compound substances have arisen.

It would thus be a vain task to try to make a rich and varied collection of minerals in certain regions of the earth, such as the vast deserts of Africa, where for many centuries there have been no plants and only a few stray animals.

It should be noted that, in the penultimate paragraph quoted here, Lamarck specified 'compound mineral substances'. In the last chapter, I stated that I did not believe that Lamarck saw base metals, such as gold or silver, as having been derived from organic matter. I believe the list of substances here cited by Lamarck shows that he had in mind rather the salts of these minerals, as well as other compound substances. At the time Lamarck was writing, there was a debate within the scientific community as to whether substances such as coal, oil and bitumen, were, or were not, of organic origin. James Hutton (1726-1797) specialized in this subject. In 1795, Hutton published a two volume work entitled *Theory of the Earth*, in which he argued that all three were of vegetable origin. Hutton was a contemporary of Lamarck's mentor, Buffon, but far more reclusive. His work, although less well known that that of Buffon, is testament to the fact that on both sides of the Channel, scientists were coming to grips with the concept of an ancient Earth, one which had changed (geologically) over time. More importantly, it was being suggested that, deep within the earth, and under the sea, there were substances of organic origin, testament to the great antiquity of life on this planet. The amount of matter now being claimed to be of organic origin, which had previously been accepted as inorganic 'rock', was substantial. It composed vast amounts of the

Earth's visible and non-visible crust. Because of the small circulation of Hutton's book, it is unlikely that Lamarck was aware of his work. Hutton's view was the minority opinion, but he was not alone and, almost certainly, Lamarck would have been aware of the debate. However, Lamarck is clearly going much further than Hutton, or any of his contemporaries, in his claims for the amount of material currently composing the Earth's crust which derived from organic material.

Lamarck's comment about the desert is interesting. It helps bring to our minds his vision of the state of the Earth's surface before life was formed – barren and desolate. As I said before, clearly inorganic matter must have existed before organic matter. Lamarck cannot have been claiming that all rock came from organic debris. However, he does seem to have been one of the first, if not the first, to grasp quite what a significant difference living organisms have made to the Earth's crust.

While other scientists seemed simply to accept that bodies were formed as the result of the reproductive process, Lamarck had clearly thought deeply on quite how plants and animals manufacture their own substance from their food, starting out as they all do from such a small and (apparently) simple beginning. Lamarck could have added that the mare, from the same grass and hay, made milk for her foal, as do all mammals, no matter how disparate their diet – vegetarian, carnivore, omnivore. The same is true of all classes of animal. They form much the same tissue and fluids as others in their class, no matter the nature of the substance(s) they ingest.

This chapter provided much food for thought.

Chapter VIII – Common faculties

I N THIS chapter, Lamarck summarized the faculties which were common to all living things and not, therefore, associated with any special organ(s). There were four:

    1) Feeding, including assimilation and repair

    2) Building the body – forming for themselves their tissues from the substances they ingested

    3) Growing, up to a certain limit

    4) The ability to reproduce

Nutrition was needed to restore losses as well as to promote growth. After growth ceased, nutrition was for the purpose of restoration and reproduction. Animals only reproduced after they had finished growing, or almost so. After the reproductive phase, nutrition, in addition to continuing to aid restoration, started to accumulate in such a way that the body became less supple, hardened, and eventually died. Assimilation (the nutrition resulting from it) always provides more solid principles or substances, than are removed or dissipated by the losses ... It is far from being true that the whole environment of living bodies tends to their destruction ... on the contrary, they only maintain their existence by means of external influences, and that the cause leading to the death of the individuals is within them and not without them.

Lamarck pointed out that, as a consequence of this constant cycle of loss and repair, after a certain period the body could not contain in it any of the molecules of which it was originally composed.

Chapter IX – Faculties peculiar to individual bodies

A s evolution has progressed, bodies have acquired certain characteristics not found in others. This chapter is largely a reiteration and expansion of the material contained in Part I.

Lamarck listed seven faculties peculiar to certain bodies, not present in others:

1) The digestion of food;

2) Respiration by a special organ;

3) The performance of acts and movements by muscular organs;

4) Feeling, or the capacity for experiencing sensations;

5) Multiplication by sexual reproduction;

6) A circulation of their essential fluids;

7) The possession of a certain degree of intelligence.

There were others, but these were the most important.

Lamarck then expanded upon his previous comments, especially the differences in respiration between that of plants, that of animals which respire through gills, imperfect and perfect respiration. Imperfect respiration was the term applied when not all of the animal's blood continually circulated through the heart. These animals had one ventricle. In perfect respiration, all blood circulated through the lungs. These animals had two ventricles.

A muscular system existed in all insects and classes beyond. Lamarck was uncertain about the radiarians. He felt there may be rudiments of muscular fibres in the sea-anemones: the coriaceous substance of their bodies makes this belief plausible, but its presence cannot be supposed in the hydra nor in most other polyps, and still less in the infusorians … although nature was able to begin the muscular system with the radiarians, yet the worms which follow them are still devoid of it.

Lamarck believed that once a characteristic or organ had been established, it would appear in all subsequent forms. If this principle is well-founded, it will confirm what I have already urged with regard to worms, viz: that they appear to constitute a special branch of the animal chain that has started afresh by spontaneous generation. The plainly marked and well-known muscular system in insects is everywhere found in animals of the following classes.

The principal of establishment was so important to Lamarck that he was prepared to consider that worms, which were possessed of a simple, but clear organization, generated spontaneously rather than consider that Nature may have experimented with rudimentary muscles in the sea-anemone but had not been able to establish them, revisiting this experiment later, first when an exo-skeleton was available for attachment, and later with animals with endo-skeletons. This surprises me.

The last faculty of which Lamarck speaks was that of intelligence, and that but briefly, since the third part of his book was going to be devoted to 'moral' characteristics, feelings, emotions and intelligence.

Section Four

PART III
Feeling and Intelligence

Introduction

THE CREATOR of All Things, in whom Lamarck believed, appears to have been a Being, who, having set Nature in motion, allowed time and evolution to progress with little or no interference.

In Part II of his book, Lamarck had concentrated upon the physical characteristics of living forms, both plant and animal. The physical forms of all plants and animals were constructed by the action of subtle (invisible) forces acting upon solid and liquid nutrients. These nutrients were reduced to a fluid and distributed throughout the body via the blood, or some other simpler fluid, released from the blood and made available by the action of the subtle fluids for transformation into tissue.

In Part III, Lamarck concentrated on the non-physical aspects of154 being – feelings, emotions, thoughts, ideas, intelligence, all of which he believed resulted from natural causes. Physical nutrition was still necessary for the building up of the nervous system and the brain, but a greater role was played by the invisible, subtle fluids.

Lamarck held that it was possible to estimate the degree of sensation present in any animal by a study of its anatomy and physiology. Plants had no organs suitable for the manifestation of the characteristics here to be studied; therefore, this Part of his book would refer only to animals.

The task which Lamarck had set himself was quite daunting:

I shall endeavor, by the use of such facts as have been collected, to determine in this third part what are the physical causes which confer on certain animals the faculty of feeling, of producing for themselves the movements which constitute their actions and, lastly, of forming ideas and of comparing these ideas, so as to obtain judgments: of performing various intelligent acts.

Lamarck acknowledged that it was impossible positively to verify his assertions but he had an 'internal certainty' as to their veracity.

If the physical and the moral have a common origin, if ideas, thought and even imagination are only natural phenomena, and therefore really dependent on organisation, then it must be chiefly the province of the zoologist, who makes a special study of organic phenomena, to investigate what ideas are, and how they are produced and preserved, in short, how memory renews them, recalls them and makes them perceptible once more; from this it is only a short way to perceiving what are thoughts themselves, for thoughts can only be invoked by ideas … it may be possible to discover how thoughts give rise to reasoning, analysis, judgments and the will to act, and how again numerous acts of thought and judgments may give birth to imagination, a faculty so fertile in the creation of ideas that it even seems to produce some which have no model in nature, although in reality they must be derived from this source …

It cannot now be doubted that the acts of the intellect are exclusively dependent on organisation …

I have devoted myself to an investigation of the only method by which nature can have brought about the phenomena in question; and it is the result of my meditations on this subject that I am now about to present.

… in every system of animal organisation, nature has but one method for making the various organs perform their appropriate functions.

There are everywhere moving fluids (some containable, others uncontainable) which act upon the organs; and there are also everywhere supple parts, which are sometimes in erethism and react on the fluids which affect them, and which are sometimes incapable of reacting; the subtle fluid moving in these parts, and modified by them in its movements, gives rise to the phenomena of feeling and intelligence as I shall endeavor to prove in this Part.

We have therefore to deal only with the relations existing between the concrete supple and containing parts of an animal, and the moving fluids (containable or uncontainable) which act on these parts.

This well-known fact has been for me as a beam of light …

Chapter I – The nervous system

THE nervous system was composed of a number of distinct parts, or systems of organs, which gradually became more complex as evolution advanced. Different animals possessed different nervous systems and, thus, experienced differently. At its earliest stage of evolution, it endowed only the characteristic of muscular movement; as evolution progressed, it allowed not only movement, but feeling. These two properties evolved separately, feeling could occur without movement and movement without feeling. Further evolution allowed the formation of ideas, of the ability to compare them and form judgments.

Lamarck referred his reader to standard texts for a full account of the anatomy of the brain, but made some basic points about its construction.

This system, wherever it occurs in animals, presents a main medullary mass, either divided into separate parts or concentrated into a single whole of varying shape, and also nervous threads which run into this mass … these organs are composed of three kinds of substances:

> 1) A very soft medullary pulp of peculiar character;
>
> 2) An aponeurotic investment which surrounds the medullary pulp;
>
> 3) A very subtle invisible fluid, which moves into the pulp without requiring any visible cavity and which is kept in at the sides by a sheath, through which it cannot pass.

Easily recognizable here are the three ingredients which Lamarck saw as being present at the very beginnings of life: a soft pulp, one which he, in a following paragraph, described as albumino-gelatinous, which was contained, an outer containing membrane and a subtle invisible fluid. It formed the centre of communication for the system of threads and cords, also with aponeurotic sheaths which penetrated the body.

One subtle fluid flowed from the centre outwards along the nerves for the purpose of exciting muscular movement; another subtle fluid flowed inwards towards the centre for the purpose of sensation.

A nucleus or centre of communication, in which the nerves terminate, is therefore absolutely necessary in order that the system may carry on any of its functions … this centre of communication is situated in some part of the main medullary mass which always constitutes the basis of the nervous system …

Lamarck then commenced to distinguish the various stages of development of the nervous system in various animals, starting with invertebrates.

… the main medullary mass … consists in some invertebrates of either a separate ganglia or of a ganglionic longitudinal cord; in the vertebrates, it forms the spinal cord and the medulla oblongata which is united to the brain.

Wherever a nervous system exists, however simple or imperfect it may be, there is always a main medullary mass in some form or other; for it constitutes the basis of the system and is essential to it … the main medullary mass … is not contained in the two hemispheres which form the bulk of what is called the brain.

… Not only can the nervous system have no

functional existence unless it is composed of a main medullary mass which contains one or more nuclei for starting muscular excitement and from which various nerves proceed to the parts … the faculty of feeling in any animal can only arise when the medullary mass contains a single nucleus or centre of communication, to which the nerves of the sensitive system travel from all parts of the body.

It may seem strange today but it was commonly held in the 18th and 19th centuries that animals had no feeling. This was the view put forward by vivisectionists who claimed, for example, that the yelps of dogs being cut up while still alive were but reflex reactions of the nervous system caused by a forcible expiration in response to the pressure of the knife. Lamarck denied this:

I am convinced that many animals possess feeling, and that some of them also have ideas and perform intelligent acts … the faculty of feeling … cannot occur unless the system possesses a single nucleus or centre of communication, in which terminates all the sensitive nerves … in animals which possess any faculty of intelligence, the nucleus for feeling is confined to some part of the base of what is called their brain; for this is the name given to the entire medullary mass … the two hemispheres, however, which are confused with the brain, should be distinguished from it; because they form together a special organ added on to the brain, have special functions of their own and do not contain a centre of communication of the sensitive system.

Lamarck did not guess the degree of 'moral' development of any animal; he made logical deductions based upon their anatomy. He continued:

... it is not in the brain properly so-called that ideas, judgments, thoughts, etc., are formed; but it is in the organ superimposed on it, consisting of the two hemispheres. Nor is it in the two hemispheres that sensations are produced ... these organs may ... undergo great degeneration without any injury to feeling or life ...

Lamarck then reminds his reader that all parts of the brain – the medulla, the hemispheres, the nerves, etc. - are covered by a membranous sheath which allows the special fluid to circulate within the systems but at the extremities of the nerves where they terminate in the parts of the body, these sheaths are open and allow the nervous fluid to communicate with the parts.

Lamarck had already explained how nutrition was liquefied and circulated throughout the body by means of the blood, or similar fluid. The more complex the blood, the more nutrition was available for the production of the nervous system. For example, mammals had more complex blood than insects and insects a more complex fluid than the watery mass in the bodies of polyps and infusorians. For Lamarck, there was a clear gradation. Æons of evolution were necessary to prepare bodily forms to receive the higher (more complex) nervous system: The medullary pulp of the nervous system, and the subtle fluid moving within it, will thus only be formed when the complexity of animal organisation has reached a sufficient development for the manufacture of these substances.

... nature ... started by producing several small masses of medullary substance when the animal organisation had advanced sufficiently to enable her to do so; she then collected them into one chief mass ... nerves cannot exist in any animal unless there is a medullary mass containing their nucleus or centre of communication

... Now it appears that the part of the mass which produced the rest is really the medulla oblongata; for it is from this part that issue the medullary appendages ... of the cerebellum and cerebrum, the spinal cord and, lastly, the nerves of the special senses.

... In ... vertebrates ... which made little use of their senses and particularly of their intellect, and which are chiefly given up to muscular movement, the brain, and especially the hemispheres should have undergone slight development, whereas the spinal cord is likely to acquire considerable dimensions. Thus fishes, which are largely confined to muscular movements, have a very large spinal cord and a correspondingly small brain.

Among the invertebrates some have a ganglionic longitudinal cord, instead of a spinal cord, throughout their length, such as the insects, arachnids, crustaceans, etc. ... the molluscs, which have only feeble supports for their muscles and generally only carry out slow movements, have no spinal cord nor longitudinal cord, and exhibit nothing more than a few scattered ganglia from which issue nervous threads.

The nervous fluid performed four functions: it instigated muscular movement, it gave rise to feelings or sensations, it produced 'inner feelings' or emotions and it gave rise to the formation of ideas, judgments, thoughts, imagination, memory, etc. Lamarck then dealt with each of these in turn.

It was not at that time understood how the nervous system worked, but it was clear that the impulses which produced muscular movement travelled from a centre in the brain, down the nerves, activating various muscles. This process was under the control of the will of the individual. A similar outward transmission took place to activate the internal organs, but this was not subject to will.

The nerves which serve for the excitation of muscular movement issue from the spinal cord in vertebrates, from the ganglionic longitudinal cord in such invertebrates as have one, and from the separate ganglia in those which have neither a spinal cord nor a ganglionic longitudinal cord. Now these nerves, destined for muscular movement, have no close connection with the sensitive system, in animals which have feeling, and when they are injured they produce spasmodic contractions and do not interfere with the system of sensations.

Lamarck was here taking the middle ground. On the one hand, he was arguing against those people who claimed that animals had no sensation and, therefore, suffered no pain. After a certain level of organization had been reached, animals did experience sensation and, therefore, pain. On the other hand, Lamarck was disagreeing with those who attributed to animals the same sensations which they, themselves, felt, considering nature cruel because many small animals, especially insects, were eaten alive. At this level of organization, Lamarck believed animals had but little capacity for sensation.

Whereas muscular activity was always generated from centre to circumference, sensation was generated in the reverse direction, from circumference to centre.

The function of the nervous system which produced inner feelings (emotions) was quite different from either of the two previously mentioned activities. This would be discussed in Chapter IV.

Not until the special accessory organs, the two wrinkled cerebral hemispheres, had evolved could there be the formation of ideas, etc. Now the accessory organ, in which are carried out functions capable of giving rise to phenomena, is only a passive organ, on account of its extreme softness; and it receives no excitation because none of its parts would be capable of reacting; but it preserves the impressions received, and those impressions modify the movements of the subtle fluids in its numerous parts. **The subtle fluids of which Lamarck was writing should not be confused with the material (cranial) fluid present in the two hemispheres.**

The nervous system is not essential to life, since all living bodies do not possess it, and it would be vainly sought among plants … The nervous system cannot then exist, nor fulfill the least of its functions, unless it consists of a medullary mass with a nucleus … this same medullary mass may at first exist without giving rise to any special sense, and it may be divided into separate parts, to each of which run nervous threads. Such appears to be the case in animals of the class of radiarians or at least in those of the division of echinoderms in which a nervous system is supposed to have been discovered … we have here the system in its greatest simplicity. It possess several centres of communication for the nerves … it does not give rise to any of the special senses, not even to sight, which is certainly the first to show itself unequivocally

… By special senses I means those which result from special organs such as sight, hearing, smell and taste; as to touch, it is a general sense, a type no doubt of all the rest, but needing no special organ

… Yet as soon as the nervous system exists, however simple it may be, it must be capable of performing some function … it will then appear very probable that the animals with the simplest nervous system derive from it the faculty of muscular movement, while yet being destitute of feeling

… Now, does this very imperfect nervous system, which is alleged to have been identified in the radiarians, also exist in the worms? I do not know; and yet there are grounds for the belief, unless the worms are a branch of the animal scale started afresh by spontaneous generation. All I know is that in animals of the class which follow the worms, the nervous system has reached a much higher stage of development, and it is quite easy to see and possess a very definite form.

Worms were a big problem for Lamarck. From one perspective, they seemed to provide a ready link between the radiarians and the insects, being of soft tissue with no skeletal structure, either internal or external. Insects had a larval form some of which resembled worms. However, there the similarity ended. Insects metamorphosed in a most extraordinary way. Nothing similar appeared among worms. Insects reproduced sexually, all being oviparous, which was not the case with worms. Furthermore, insects did not reproduce in their larval state, but only after metamorphosis. Insects exhibited the earliest traces of hard material in their organization such as could support muscles with their constant contractions. These muscles were extremely fine as were the nerve fibres which activated them. Many insects had developed wings, found in no previous form. It would assist Lamarck's argument if the rudiments of a nervous system, muscular activity and oviparous sexual reproduction could be found among the worms, but he was not convinced that it could. I am unclear quite why Lamarck should have thought that the problem would be solved more easily if worms were spontaneously generated, since the main distinctions seems to be between worms and insects, rather than worms and the radiarians.

Indeed … the first appearance of the nervous system has hitherto seemed to be in insects. … in all the animals in this class it is very clearly defined, and presents a ganglionic longitudinal cord … This longitudinal cord, which ends anteriorly in a subbilobate ganglion, constitutes the main medullary mass of the system, and from its ganglia … nervous threads proceed to the various parts of the body.

The subbilobate ganglion … has to be distinguished from the other ganglia of the cord, since it gives rise directly to a special sense – that of sight. This terminal ganglion is, then, really a small and very imperfect brain, and doubtless contains the centre of communication of the sensitive nerves, since the optic nerve runs into it.

… in the insects, the nervous system begins to present a brain and a single centre of communication for the production of feeling … they suffice at least to constitute feeling, although incapable of producing ideas.

Referring back to my earlier comments about cruelty in nature, insects being eaten alive would feel something, if only pressure. However, since they have no ability to think, or produce an idea, they would not be able to experience fear, as we understand it.

The state of the nervous system … in insects … is almost the same in the five following classes viz: arachnoids, crustaceans, annelids, cirrhipedes and molluscs

… When nature had supplied the nervous system with a true brain … she … started the rudiments of the senses of hearing in the crustaceans and molluscs. But it still continues to be a very simple brain, appearing to be the basis of the organ of feeling, since the sensitive nerves and those of the existing special senses proceed to unite with it.

… the terminal ganglion … to some extent bilobate, still shows no trace of the two wrinkled hemispheres, so susceptible of development, which in the most perfect animals cover over the true brain … hence the function, for which these new accessory organs are adapted cannot be performed in any of the invertebrates

… It is only among the vertebrates … (viz: the fishes) that she [nature] started the rudiments of the accessory organs … which consists of two wrinkled hemispheres … as a rule the name of brain is given to the entire medullary mass enclosed within the cavity of the cranium. We should, however, distinguish between the brain properly so called and its accessory organ … for the accessory organ fulfills altogether special functions, and it is neither essential to the brain nor even to the maintenance of life. It therefore deserves a special name and I propose to call it the hypocephalon.

Now this hypocephalon is the special organ in which ideas and all acts of intelligence are carried out.

Chapter II – The Nervous Fluid

LAMARCK argued that only a subtle, i.e. invisible, fluid could produce the observed effects of the nervous system. It was because this fluid could not be directly observed in the same way as material fluids that it had not been studied as it should have been. He maintained that the existence of subtle fluids was now well established and accepted. He had earlier mentioned gravity, magnetism and electricity; in this chapter he included galvanism in the list, although he added that this was a form of electricity.

While Lamarck allowed that other, as yet unknown, subtle fluids might exist, he stated that he was convinced that it is electric fluid, which has been modified in the animal economy and to some extent animalized by its residence in the blood, and which has there undergone sufficient change to have become containable and to remain entirely within the medullary substance of the nerve and brain, to which it is incessantly provided by the blood.

Lamarck had reached this conclusion because all visible fluids, such as blood, lymph, etc., in any animal's body moved far too slowly to account for muscular activity. Furthermore, fluids moved in one direction only; they could not start and stop their movement with the extreme rapidity which was necessary for any animal's complicated, and often sudden, movements. Not only did this fluid move with extreme rapidity (Lamarck suggested approaching the speed of light), it did not form any visible channels.

The nervous fluid is … quite distinct from the ordinary electric fluid, since the latter passes through every part of our body at its usual velocity and without any pause …

Lamarck reminded his reader that every tissue in any animal body had been introduced by means of fluid nutrition. As evolution progressed, nutrition ingested by way of the mouth was not merely absorbed into the tissues, it was digested, after having been reduced to a liquid in the gut. It was then transferred to the blood stream, or equivalent in less complex animals, for distribution throughout the body. Oxygen, removed from water or air, was similarly taken up by the blood, or equivalent substance, for distribution throughout the body. Lamarck suggested that a form of electricity also entered the body, firstly by means of food, but possibly also during respiration. This 'electricity' became animalised, that is converted into a state in which it did no harm to the body but could be contained within the sheaths surrounding the brain and nerves, allowing it to flow to the muscles for both voluntary and involuntary movement.

The nervous fluid of cold-blooded animals, being less animalised, is more allied to the ordinary electric fluid … This is why our galvanic experiments produce very energetic effects on the tissues of cold-blooded animals like frogs; and also why in certain fishes such as the torpedo, the electric eel, and the trembling catfish, a large electrical organ generates electricity which is completely adapted to the animal's needs.

Lamarck felt that since this special form of electric fluid was constantly being used by the body, by the internal organs even at rest, it was necessary for it constantly to be replaced, which was achieved by the constant flow of the nutrition-bearing blood, or other fluid.

The nervous fluid in the brain does not merely convey sensations from their nucleus, and move about in various ways, but it also stamps impressions on the organ, and these impressions last longer or shorter according to their depth. **Lamarck indicated that he would return to this topic later.**

All parts of the nervous fluid are in communication, in the system or organs which contain them … either a part of the fluid may be set in motion, or nearly the whole fluid … the nervous influence which moves the muscles only demands a simple emission of a part of the nervous fluid towards the muscles which have to act …

In acts of intelligence, the organ of understanding is only passive, and is prevented from reacting by it extreme softness; it acquires no activity from the nervous fluid but merely impressions, of which it preserves the tracings; the part of the fluid, which works in the various portions of this organ, is modified in its movements by the tracings already present, and at the same time traces more; so that the organ of understanding, which has only a narrow channel of communication with the rest of the nervous system, uses only a part of the whole fluid of the system.

… it appears probable that the entire mass of nervous fluid secreted and contained in the system is not at the disposal of the inner feeling of the individual, but that some of it is held in reserve to provide for the continuance of the vital functions … since the nervous fluid is never used without a proportional amount of loss, it follows necessarily that the individual is only free to use up a certain part of it: untoward effects

ensue even when this part is run too low, for
then some of that held in reserve becomes
available and the vital functions suffer
accordingly.

It would seem that here Lamarck was referring to what is
commonly referred to as 'shock', when a great emotional
upheaval uses up all the resources of the nervous system,
resulting in the collapse, or near collapse, of the victim.

Chapter III – The mechanism of sensation

L AMARCK had established that sensation, as distinct from irritability, first became possible in insects with the appearance, however rudimentary, of the true brain, the medullary mass. At its first appearance, it is likely that nothing more than a sensation of pressure would be possible, without any feeling of actual pain. This circumstances is one with which many people will be familiar from having received treatment by a dentist! As the medullary mass grew, the ability to perceive sensations increased, finally reaching its acme in the human body. Higher animals were able to feel more pain than lower and humans were capable of feeling more pain than other mammals, but the process was gradual.

In this chapter, Lamarck did not consider the process of evolution itself. He considered how sensations were felt, the physiology of the process. He commenced by stating that it was an illusion, or hallucination as he termed it, that sensitivity (pain) occurred in peripheral parts of the body. The perception of pain was a function of the medullary mass. At this point of his book, Lamarck was only considering physical sensibility. Emotional (moral) sensibility, also a function of the medullary mass would be addressed in the next chapter.

Physical sensibility could be experienced in the peripheral areas of the body, or in the internal organs if they were disordered.

A large number of simple nerves which start from all the sensitive parts of the body … proceed to their destination in the nucleus of sensations [which is] the centre of communication [in the medullary mass]

… every impression, internal or external, that

any individual receives, immediately causes an
agitation throughout the system or the subtle
fluid contained in it, and consequently
throughout the entire body, although it may
pass unperceived … this … gives rise to a
reaction which is brought back from all parts
to the common nucleus, and there sets up a
singular effect … which is thereafter
propagated through the one nerve, that does not
react, to the point of the body which was
originally affected.

Lamarck repeats this description several more times in slightly
different words. It is more complicated than an impulse travelling
from the periphery to the centre as the result of some agitating
action and then rebounding down the nerve to the point of origin,
giving the perception of sensation at the periphery. Lamarck
believed that, because all of the nervous system throughout the
body was interconnected, both directly and via the medullary
mass, the effect of the 'shock' which precipitated the action was
felt throughout the whole body. Where there was action, there
was reaction. The impulse then returned to the medullary mass
and from thence travelled back down the nerve originally
affected. Lamarck stated that this nerve was the one that was not
at that time 'reacting' because he saw its fluid impulse as having
been expended, leaving the nerve quiescent. It was reactivated by
the returning impulse, which travelled its length to its origin, or
apparent origin. Lamarck mentioned the phenomenon of
phantom pain in an amputated limb.

In the case of animals which have a spinal
cord, there starts from every part of their
body, the most deeply situated as well as the
most superficial, nervous threads of extreme
fineness, which without any division or
anastomosis proceed to the nucleus of
sensations. Now these threads, in spite of

junctions that they form with others, travel
without any discontinuity to their nucleus,
always retaining their individual sheaths …
Each nervous thread may thus be distinguished
by the name of the part where it starts, for it
only transmits impressions made on that part. …
the nerve which brought in the original
impression, and thus set up the agitation of
fluid in all the rest, is the only one which
gives no reaction; for it is the only one which
is active while all the rest are passive … the
whole effect of the shock … must therefore be
carried off by this active nerve.

Lamarck's use of 'active' and 'passive' is somewhat confusing. If I
am correct, Lamarck was suggesting that the impacted nerve was
active, while all the others were passive, until the impulse reached
points of connection, either to other nerves or to the controlling
centre. The fluid impulse in the originally affected nerve being
expended, this nerve becomes passive, while all the other nerves,
and the medullary mass, become active, although individually at a
lesser intensity. Impulses return via these activated nerves to the
original, now passive nerve, which becomes re-activated, while
the rest become passive once again. It was this second bout of
activity in the originally affected nerve which produced sensation.

Lamarck re-expressed his hypothesis, which I will repeat here.
Hopefully, the reader will be able to understand what is a
somewhat confusing proposition.

… the nerve at whose extremity the original
impression was received is the only one which
does not subsequently react; and that for this
reason the general reaction from the other
nerves of the system, on reaching the common
nucleus, is necessarily transmitted into the
only nerve that is at the moment in a passive
state, and thus conveys to the point first

affected the general effect of the system, that is to say sensation.

All of this occurs with extreme rapidity, such rapidity that there is no awareness of any time delay.

We carry out almost simultaneously by means of our organs two acts of essentially different kinds, one which makes us feel while the other makes us think … they cannot have a common origin.

A special system of organs is certainly needed for producing the phenomenon of feeling, since this is a faculty peculiar to certain animals and not general for all. So too a special system of organs is necessary for carrying out acts of the understanding; for thought, comparison, judgment, reasoning are organic acts of a very different character from those producing feeling … since all sensation arises from a special sense, it follows that the consciousness of one's thoughts is not a sensation, but differs radically from it, and must be kept distinct … we may … think without feeling, and feel without thinking … for each of these two faculties there is a separate system or organs.

… every simple idea arises exclusively from a sensation. But I hope to show that not every sensation produces an idea … for the production and impression of a permanent idea a special organ is needed

… It is a long way from a simple perception to an impressed and permanent idea. Indeed no sensation, which causes only a simple perception, makes any impression on the organ;

it does not need the essential condition of attention … memory, whose seat can only be in the organ where ideas are traced, can never bring back a perception which did not penetrate this organ, and therefore left no impression on it.

I regard perceptions as imperfect ideas, always simple, not graven on the organ and needing no condition for their occurrence … these perceptions, by means of habitual repetitions which cut out certain channels for the nervous fluid, may give rise to actions which resemble those of memory.

… whenever an individual notices sensation, identifies it, and distinguishes the point of its body on which it takes effect, the individual then has an idea, thinks, carries out an act of intelligence, and must therefore possess the special organ for producing it.

… when the system of sensations exists without the system of understanding, the individual performs no act of intelligence, has no ideas, and only derives from its senses simple perceptions which it does not notice, although they may arouse its inner feeling and make it act.

Clearly, the 'individual' referred to in these last two paragraphs was not necessarily human. Many animals are capable of identifying sensations and distinguishing the point on their body to which they relate. Lamarck held that all of these individual animals were capable of thought and of carrying out acts of intelligence, however 'low' or simple that intelligence might be. This illustrates a far greater understanding of, and respect for, animals and their inherent mental capabilities, than was generally held by others at that time.

Chapter IV – Feelings, emotions and actions

LAMARCK stating that nobody, so far as he knew, that had considered what, to him, was the interesting subject of that singular faculty, with which certain animals and man himself are started by endowed consisting in the capacity to experience inner emotions called forth by the needs and various causes external or internal; from this faculty arises the power of performing diverse actions.

Lamarck was here moving from the physical sciences of physics, chemistry and biology with which he had thus far been concerned, to the, as yet unnamed, science of psychology. He would be considering, not merely the physical functions of the nervous system which underlie the phenomena of movement and sensation, but how some animals and humans have the power, as a result of 'moral' activity, to change (within limits) their actions, to use the power of their thought and their will to perform diverse actions.

… that portion of the nervous fluid which serves for the excitation of muscles independently of the individual, and often also that portion contained in the hemispheres of the brain, are sheltered from the agitations which constitute emotions.

Lamarck explained that there were two distinct types of movement, or agitation, which affected the cerebral mass: partial agitations which become general and end by reactions; it is this kind of agitation which produces feeling … agitations which are general from the first and form no reaction; it is these which constitute the inner emotions, and it is exclusively with these that we shall now deal.

First it was necessary to consider something which Lamarck described as 'the feeling of existence' for which he would use the term 'inner feeling' … a very obscure feeling possessed by animals whose nervous system is sufficiently developed. This feeling … is the origin of the inner emotions and consequently … enables individuals to produce for themselves the movements and actions which their needs demand. Plants do not have such a capacity. Flower petals may open and close, but they do so in response to light; the plant does not have the capacity to choose for itself when its petals will open or close. The same situation applies to the simplest forms of life, the infusorians, the polyps and the radiarians. Insects, however, do have a measure of control over their actions. They may decide the exact moment when they will spread their wings and take flight. The need which demands this action may be that of the search for food, or a mate, or to escape from a predator, but their wings only spread in response to a muscular stimulus initiated in the insect's brain. All animals, including humans, react to a sudden stimulus without thought, as Lamarck would later discuss.

All animals with an adequate nervous system are constantly receiving impressions throughout their sensitive parts … these impressions are very weak, although they vary in intensity according to the health of the individual … Yet the sum total of these impressions, and the confused sensations resulting from them, constitute in all animals subject to them a very obscure but real inner feeling that has been called the feeling of existence … This … feeling … is general, in that all the sensitive parts of the body share it. It constitutes that ego with

which all animals that are merely sensitive are
imbued without knowing it, while those which
also possess an organ of intelligence may
notice it.

Even when completely at rest, we still experience this inner feeling of existence so long as we are awake. When 'aroused', this inner feeling is the source of our activity. In humans, most activity is under the direction of our thought, our will. In animals with a less sophisticated cerebral mass, a greater proportion of activity is automatic, or instinctive, but all animals with even the rudiments of brain development, from insects onwards, experience some degree of control emanating from this inner feeling.

Only humans have the capacity to evaluate their inner feelings; animals merely experience them. Nevertheless, even some animals have the capacity to evaluate a situation and make a decision as to the most appropriate or most necessary action. This is only the case when the emotions are weak or moderate; when the stimulus is sudden or powerful, both humans and animals react without thinking.

Humans, and some higher animals, are moved by the sight, or thought, of something: a precipice, or a tragic scene, either real or on the stage or even on a picture, etc., etc.: and where is the power of a fine piece of music well executed, if not in producing emotions of our inner feeling? Consider again the joy or sorrow that we suddenly feel on hearing good or bad news ... what is it but the emotions of that inner feeling, which is very difficult to master on the spot?

... Our habits, temperament, and even education,
modify this faculty of undergoing emotion

... Moral sensibility is very different from ...
physical sensibility ... the latter only arises
from impressions produced on our senses ... these
can likewise stir our inner feeling ... moral
sensibility is nothing more than a very

delicate susceptibility to emotion … it is the source of humanity, kindness, friendship, honour, etc. Sometimes, however, circumstances make this quality almost as baneful to ourselves as its presence in others in beneficial … A good education shows us the necessity on innumerable occasions for repressing our sensibility up to a certain point … hence results that decorum and amenity in the expressions used in conversation, in short, that careful restraint in the expression of ideas which gives pleasure without ever wounding, and confers a quality of high distinction to those who possess it… But the limits are sometimes passed … many men possess certain propensities which lead them constantly to resort to dissimilation … [they] habitually … hide their thoughts and such of their actions as may lead to the end they have in view.

Now since every faculty, that is not used, gradually degenerates until it almost becomes extinct the moral sensibility which we are here discussing is almost absent in them; and they do not even esteem it in persons who still possess it …

Lamarck is holding true to his conviction that all life, both physical and moral, is initiated by the same subtle forces and bound by the same laws. He had earlier argued that use increased development and disuse caused degradation of physical parts; he now applies this same law to moral characteristics. Higher ethical attributes need to be practiced; unless they are practiced, they will become degraded and society's moral fibre will degenerate as surely as does physical fibre. Lamarck's expansion of the laws of physical evolution beyond personal psychology into the arena of social interactions would today be referred to as social evolution or social psychology. There was no such discipline when Lamarck was writing.

The emotions of the inner feeling were of two types: moral emotions initiated by thought and physical emotions derived from sensation.

A moral emotion when very powerful may temporarily extinguish physical feeling, disturb the ideas and thoughts, and cause some enfeeblement in the functions of several of the most essential organs … distressing news, when unexpected … extreme joy, produces emotions whose consequences may be of this kind.

It is also known that among the minor effects of these emotions are digestive troubles or pains; and that, in the case of elderly people, when the emotions are at all strong, they may be dangerous and even fatal … Fanatics, for instance, are people whose moral feeling is so exalted as to overcome the impressions of the tortures, which they are forced to undergo … when very strong [moral emotions] may also disturb the intellectual faculties

… I shall conclude these remarks with a reflection that I believe to be well founded, viz: that the moral feeling exercises in course of time a greater influence on the organisation than the physical feeling is capable of working … constant and real griefs in any individual may set up degeneration of the abdominal viscera; and that these degenerations, once started, may in their own turn propagate the inclination to melancholy, even when there is no longer any real cause for it.

Lamarck then applied the same principle to the inheritance of moral characteristics that he had to the inheritance of physical characteristics:

Reproduction, indeed, may transmit a tendency of the organs or a state of the viscera adapted for giving rise to any special temperament, inclination or characteristic; but it is essential that circumstances should favour the development of this tendency in the new individual; for otherwise the individual would acquire another temperament, inclinations, and characteristics. It is only in animals of very low intelligence that reproduction transmits almost without variation the organisations, inclinations, habits and special peculiarities of each race.

In the same way that physical characteristics stabilized in a living population so long as that population's external environment remained the same, so, too, would similar moral characteristics appear, and re-appear in subsequent generations, so long as those moral characteristics served their purposes. The external characteristics which might bring about change in moral characteristics were more complex than those which might bring about physical change, because they included the surrounding moral environment, both that of other individuals in close proximity as well as society itself. Change, in both instances, was a response. Lamarck then acknowledged that this was not the appropriate forum for further exploration of these ideas. He returned to his former topic.

... when the emotions in question are so powerful as to cause an agitation in the nervous fluid sufficient to affect the movements of the portion contained in the cerebral hemispheres and also that which controls the involuntary muscles, the individual then loses consciousness, and suffers from syncope; and the functions of his vital organs are more or less deranged.

… If sleep is imperfect … the inner feeling (being suspended) no longer directs the movements of the fluid in the nerves, and the individual is then abandoned to dreams, that is, to involuntary recurrences of his ideas following one another in characteristic disorder and confusion.

… it seems to me obvious that the inner feeling is the only factor that produces actions in man and such animals as possess it; that this feeling only works when prompted by its emotions; that it is moved sometimes by acts of intelligence, sometimes by a need or sensation acting suddenly upon it; that in men, its weak emotions may be controlled since intelligence is highly developed, but that in animals this can only be done with great difficulty and never in those that lack intelligence; that its functions are suspended during sleep … its functions may also be interrupted and disturbed during the waking state … it is the product … of the individual's feeling of existence, and … of the harmony in the parts of the nervous system; as a result of which the free portions of the subtle nervous fluid are all in communication and capable of undergoing a general agitation.

… it also appears to me obvious that moral sensibility only differs from physical sensibility, in that the former results exclusively from emotions prompted by acts of the intelligence; while the latter is produced only by emotions aroused by sensations and the needs which they evoke.

… the intimate feeling of existence possessed by animals which have the faculty of feeling but not that of intelligence, confers upon them at the same time an inner power, which only works through the emotions called up by the harmony of the nervous system, and which causes them to carry out actions without any co-operation of the will. But such animals as possess both the faculty of feeling and that of intelligence, have this advantage over the first: that the inner power, which inspires their actions, is susceptible of receiving its driving emotions either through sensations produced by internal impressions and wants that are felt, or through a will which, though more or less dependent, is always the result of some act of intelligence.

Chapter 9

PART III
Feeling and Intelligence
(continued)

Chapter V – The Force which produces action

THIS chapter is a recapitulation of Lamarck's arguments thus far. It is as if Lamarck, having built his edifice with such care, felt the need to make sure its foundations were secure before continuing.

Plants can satisfy their needs without changing their position … The case is different with animals: for, with the exception of the most imperfect animals at the beginning of the animal chain, the food on which they live is not always at hand, and they are obliged to carry out movements and actions in order to procure it.

… nature has wrought her various productions by slow and gradual stages … we feel that she must have begun by borrowing from without, that is, from the environment, the force which produces the organic movements and those of the external parts; that she afterwards transferred that force within the animal itself, and that, finally, in the most perfect animals, she made a great part of that internal forc available to their will

… animals that do not yet possess a nervous system cannot contain within themselves the force which produces their movements … this force must be outside them

… This force is the result of subtle fluids (such as caloric, electricity, and perhaps others) which incessantly penetrate these animals from the environment, set in motion the visible and contained fluids of their bodies, and by exciting the irritability of their containing parts, give rise to the various movements of contraction which they produce.

… these subtle fluids … carve out special routes … which, when continued or repeated, give rise to habits.

… The most imperfect animals … only feed by means of absorption … organisation develops as life goes on … new routes must have been cut out … increased in number and progressively varied for the furtherance of the movements of contraction

… Such … is the cause of movements in the most imperfect animals; movements that we are lead to attribute to their own initiative, and to regard as a result of their faculties, because we know that in other animals the source of them is within their bodies

… If nature had confined herself to her original method … animals would have been simply passive machines and nature would never have produced … sensibility, the intimate feeling of existence, the power of acting and, lastly, ideas by means of which she has created the most astonishing of all, viz. thought or intelligence.

… she gradually prepared the way by increasing the coherence of the internal parts of animals, by diversifying their organs, and by multiplying and compounding their contained fluids, etc.; thereafter she was able to transfer into the interior of these animals that force productive of movements and actions.

… when [nature] … establish[ed] a nervous system … as among insects … animals were endowed with the intimate feeling of their existence; and henceforward the force productive of movements was transferred into the interior of the animal itself.

… every need that is felt produces an emotion in the individual's inner feeling … the inner feeling then takes the place of will … every animal which does not possess the special organ … by means of which thoughts, judgments, etc., are produced, has really no will

The case is very different in animals which nature has endowed with an additional special organ (two wrinkled hemispheres surrounding the brain) … These animals control their power of acting in proportion to the perfection of their organ of intelligence; and though they are still strongly subordinated to … habits … they yet enjoy a will that is more or less free; they can choose, and introduce variation into their actions or at least into some of them.

The nervous fluid, which is constantly being formed throughout life, is as constantly being consumed by the use which the individual makes of it. One part of this fluid is kept occupied … in the maintenance of vital movements and of the functions necessary to life. The other part of the fluid is at the service of the individual … either for … movements, or for the performance of … acts of intelligence.

The individual thus uses up its invisible fluid … and it would exhaust the whole of the available part, if it continued too long … Hence arises the need of rest after a certain period of action … when the individual uses up too much of the free fluid … the functions of its vital organs suffer … these organs flag to some extent

… It is chiefly man … that uses up his physical strength in this way; for, of all his actions, those that use the most nervous fluid are the prolonged acts of his understanding, thoughts, meditations, and, in short, the continued efforts of his intellect … the physical strength is proportionally diminished

… Every action is caused by some movement in the fluid of the nerves. Now when this action has been several times repeated there is no doubt that the fluid cuts out a route which becomes especially easy for it to traverse … the power of habit over actions is inversely proportional to the intelligence of the individual, and to the development of his faculty of thinking, reflecting, combining his ideas, and varying his actions

… Animals which are only sensitive … have nothing but perceptions (often very confused), do not reason, and can make little variation in their actions. They are therefore permanently subject to the power of habit.

Thus insects … experience perceptions of the objects which affect them, and seem to possess a memory gained through a repetition of these perceptions. They can, however, neither vary their actions nor alter their habits, since they possess no organ to give them this power.

… in all such animals that are devoid of intelligence, the tendencies of action can never be the product of a reasoned choice or judgment or experience … the causes which stimulate actions, are sometimes internal and sometimes external … our impressions can only be internal; and the sensations of external objects, a reasoned choice or judgment or experience … derived from our special senses, cannot produce in us any but internal impressions

… This propensity of animals to the preservation of habits, and to the repetition of the resulting actions … is propagated to succeeding individuals by reproduction so as to preserve the new type of organisations and arrangements of the parts; thus the same propensity exists in new individuals, before they have even begun to be succeeding individuals by reproduction so as to preserve the new type of organisations and arrangements of the parts; thus the same propensity exists in new individuals, before they have even begun to exert it.

Hence it is that the same habits are handed on from generation to generation in the various species or races of animals without any notable variation so long as no alteration occurs in their environment.

Lamarck acknowledged the role reproduction played in establishing altered habits or characteristics of organisation – provided there was no alteration in their environment, physical or moral. External influences, physical or subtle, drive evolution; change in living bodies, plant or animal, was always a reaction to external factors.

It is only among the vertebrates, and particularly among the birds and mammals, that we find the characteristics of a true skill; for in difficult situations their intelligence may assist them to vary their actions notwithstanding their subjections to habit. These characteristics however are not common and it is only in a few races that we often witness them.

There is nothing really new in this chapter, which functions mostly as a recapitulation, a bringing together of Lamarck's arguments thus far.

Chapter VI – Of the Will

This chapter is comparatively short and serves as a bridge between the consideration of the effect of the subtle, invisible, life force (aura vitalis) within organic bodies and how the matter which constitutes physical bodies reacts to this force and the 'matter' of our mind and intellect, of what it is formed and how it reacts to the subtle forces which activate it.

In this chapter Lamarck considers the will in both humans and animals.

… the will is the immediate result of an active intelligence; for it is always the effect of a judgment and hence of an idea, thought, comparison or choice … an operation of the organ of understanding, to perform some action, combined with the faculty of exciting an emotion of the inner feeling which can produce that action.

… animals which have no organ for intelligence cannot carry out any acts of will … There are … several different sources from which the actions of animals may be derived … in animals which have no organ for the will, the inner feeling can only be moved by means of sensations; whereas in those which have an organ for intelligence, the emotions of the inner feeling may either be the exclusive result of sensations, or that of a will born from an operation of the understanding.

… we have three distinct sources for the action of animals … external causes which excite …

irritability … inner feeling moved by
sensations … inner feeling moved by will.

Lamarck repeated his previous claim that animals of the first group (infusoria, radiarians and polyps) moved only as the result of a reaction to external stimulus without the use of muscles and that the second group (all other invertebrates) were able to make muscular movements in response to sensations but not from any idea or judgment. Even animals endowed with two wrinkled hemispheres mostly acted as the result of external stimulus, such as sight or sound, without judgment, although they had a limited ability to form judgments and make some choice in regard to their actions.

Even humans performed acts without thought, such as when they picked up something hot. They dropped it immediately, without conscious thought.

Although invertebrates from insects upwards were capable of inner feeling, they were not capable of acts of will. It was only animals with wrinkled hemispheres which were capable of forming complex ideas, thoughts, comparisons, judgments, etc. Fish and reptiles had 'imperfect brains' which did not fill their cranial cavity, indicating very low intellectual ability. Only birds and mammals were possessed of complete brains and were able to carry out acts of intelligence. Birds and mammals also possessed memory which was formed by the nervous fluid making traces in the soft material of the hemispheres and the brain. Random agitation of the nervous fluid over these traces during sleep was the cause of dreams which were known to occur in some animals as well as in humans.

Although possessed of the ability to register thoughts and to make decision, for the most part animals made little use of these faculties. Most of their actions were based on instinct in response to sensations and for this reason they rarely made mistakes. Because animals lead very similar lives, the traces made by the nervous fluid were very similar in each member of a population, or species. For this reason, all members of a population or species tended to act in a similar manner, not displaying much individuality.

Lamarck then considered the will in relation to humans.

Since the will is always dependent on some judgment it is never really free … the produce of a judgment must vary in different individuals for the reason that the elements which enter into the formation of this judgment are apt to be quite different in different individuals

… In fact, so many and various elements enter into the formation of our judgments, so many are present which ought not to be present … or are affected by our disposition, health, age, sex, habits, propensities, state of our knowledge, etc., that the union of these elements gives rise to very different judgments on the same subject in different individuals. The fact that our judgments depend on so many inappreciable elements has given rise to the belief that our determinations are free, although in reality they are not so, seeing that the judgments that produce them are not so …

One of the reasons there were so many different opinions was that the information upon which the judgments were formed was incorrect. Another reason was the great inequality in intellect between various persons, something not apparent in animals.

… while man derives great advantages from his highly developed intellectual faculties, the human species in general suffers from them at the same time considerable disadvantages; since these faculties confer the means for doing harm as easily as good, and their general effect is always to the disadvantage of those individuals who make least use of their intelligence, and this is necessarily the case of the greater number … the main evil in this respect resides in the extreme inequality of intelligence between individuals … the thing most important for the improvement and happiness of man is to diminish as far as possible this enormous inequality, since it is the origin of most of the evils to which he is exposed.

It could be argued that most of the wars and persecutions which mankind has suffered during recorded history were instigated by the ruling classes, both civil and religious, rather than the working poor, who were generally considered to possess lower intelligence than the 'Upper', ruling classes, sometimes known as the 'intelligensia'. France had, not long before, suffered its Revolution, as the result of which thousands died. This Revolution had been led by members of the Lower classes and was directed against the aristocracy, of which Lamarck was a minor part. Perhaps Lamarck was thinking of this? Bearing in mind the political climate of the times, these final comments are unlikely to have been received favourably by many of Lamarck's contemporaries.

Chapter VII – Of the Understanding

LAMARCK considered 'understanding' to be the most difficult of the subjects he was studying in these final chapters. The question is, how purely physical causes, that is to say, simple relations between different kinds of matter, can produce what we call ideas; how, out of simple or direct ideas, complex ideas may be formed; how, in short, out of ideas of any kind, faculties can arise, so astonishing as those of thought, judgment, analysis and reasoning.

... I start with the conviction that all acts of intelligence are natural phenomena and hence derive their source from physical causes.

In the earlier part of his work, Lamarck had been at pains to draw attention to the difference between solid matter and subtle (invisible) forces, such as gravity, electricity, magnetism and the aura vitalis. It was the action of this last subtle fluid which caused the differentiation between inorganic and organic matter. Now Lamarck was stressing that, since all of the before-mentioned subtle fluids were part of nature, they should all come under the heading of 'physical cause'.

Under the term 'understanding', Lamarck included all intellectual faculties: the forming of ideas, comparing, judging, thinking, analyzing, reasoning and memory.

Lamarck held that all of these activities took place in the hypocephalon, parts of which showed definite differentiation.

All intellectual acts whatever originate from ideas, acquired either at the time or previously ... Every idea of any kind originates either directly or indirectly from a sensation.

Lamarck quoted Locke: *That there is nothing in the understanding which was not previously in sensation.* Lamarck seemed also to embrace another idea of Locke's, that each individual life started du neuvo at conception, that the brain/mind was a tabula rasa, a blank slate, upon which impressions were engraved. Eastern philosophers accepted the pre-existence of the soul, thus allowing some impressions and memories from previous life experiences to be impressed on the brain during growth before birth. Lamarck, in accordance with Western philosophy, did not consider this possibility. His assertions rested on the assumption that the hypocephalon received impressions only from birth onwards.

I lay down, then, as a fundamental principle and unquestionable truth, the proposition that there are no innate ideas, but that all ideas whatever spring either directly or indirectly from sensations which are felt and noticed.

The soft tissue of the medullary pulp and the hypocephalon were passive; they merely provided the means for the nervous fluid to execute its function. The medullary parts of which the hypocephalon consists, receive and preserve the traces of all the impressions made upon them by the movements of the nervous fluid ... the only active element ... is the nervous fluid itself

... The medullary mass ... consists ... of an inconceivable number of separate and distinct parts, from which result a vast quantity of cavities of infinitely varied size and shape and appearing to occupy distinct regions, equal in number to the intellectual faculties of the individual ... it may come about, the composition of the organ is different in each region, for each is devoted to some individual faculty of the intelligence.

Today, the different functions of the various parts of the two hemispheres are well mapped.

The medullary part of the hypocephalon contained numerous fine white fibres, but these fibres were not associated with movement, as were other nerve fibres in the body ... they are so many individual canals, each terminating in a cavity, which would be in the form of a cul-de-sac unless they communicate together by lateral paths. These cavities ... are as innumerable as the tubular threads leading to them; and it may be presumed that it is on their internal walls that the impressions brought by the nervous fluid are engraved

... the organ of intelligence, like the rest, develops according to its use. The same applies with each special kind of intellectual faculty ... It is therefore not true that any of our intellectual faculties are innate ... These faculties and propensities grow and strengthen according to the exercise which we give to their underlying organs ... unless we exercise these faculties and inclinations ourselves we gradually lose the aptitude for them ...

Lamarck believed that, although we all seemed so different from one another, each with our own thoughts and ideas, nevertheless we had but small control over our development, especially as a child, since we had so little control over the impressions to which we were subjected. No two humans were born to identical circumstances or underwent identical experiences. While some children were fortunate enough to be brought up in happy circumstances, others were subjected to environments, both material and personal, highly disadvantageous to us by their mode of life, thought and feeling; and sometimes by illadvised weakness they spoil us and let us acquire many pernicious faults and

habits whose consequences they do not foresee …
It is difficult to conceive how great is the
influence of early habit and inclinations on
the propensities which will some day dominate
us, and on the character which we form.

Lamarck asserted that behaviours acquired in childhood were not easily changed in later life, especially if the channels formed by the nervous fluid had made a deep impression. As we aged, the soft pulp of the wrinkled hemispheres hardened slightly. Impression were made less easily. Memory became impaired. It was easier to recall events long past, whose impressions were more deeply etched. Lamarck then considered ideas. He stated there were two types: simple or direct and complex or indirect. Each of these also occurred as two subtypes: physico-moral ideas which were clear and vivid and moral ideas which were vague and unclear. Physico-moral ideas were the result of impulses received from our sense of sight, hearing, etc. These were clear and vivid. Moral ideas were memories of things previously seen or heard, or mental conversations, which we see with our inner eye, or hear with our inner ear, experience by our inner feeling, in a way which is difficult to describe, which is as real as the physico-moral idea, but totally different.

A simple idea, originating as the result of a sensation, can only be formed if the sensation is noted. It is then impressed upon the soft tissue of the brain and is available for recall. A complex idea is formed when two simple ideas, perhaps one new and one already stored in the memory, are combined. This first stage complex idea may then become a second stage complex idea by being associated with yet another new idea or with some other already formed complex idea. Thus there was no limit to the number of complex ideas which could be formed by any individual. The individual who habitually exerted his intellect might continually form new ideas. Some animals were capable of receiving impressions, of forming ideas and of making judgments, but none did so to the extent that did humans.

Chapter VIII – Principal acts of understanding

LAMARCK identified four principal acts of the understanding: attention, thought, memory and judgments. Desire was not an act of the understanding; it was the drive to satisfy a physical or moral need. Physical needs arose from physical sensation, hunger, thirst, pain, etc., while moral needs arose from thoughts, to escape pain or hardship, to achieve comfort or escape danger, etc. These aroused the individual's inner feeling and prompted the desire for188 action but were not acts of understanding. All animals experienced physical needs and those above the polyps were able to move to satisfy them; many animals experienced moral needs and could make simple decisions to satisfy them, but few animals had the capacity for understanding encompassing all four necessary ingredients and only humans to any great degree.

This chapter concentrated on human understanding.

Lamarck considered first the act of attention which he saw as an act of the inner feeling which prepared the organ of thought for carrying out its function. The effort directed nervous fluid towards the appropriate part, preparing it to re-awaken an idea already impressed into its substance or to receive a new idea. Unless attention was used to focus the nervous fluid, no impression was registered from sensations received. A multitude of sights and sounds passed us by, unnoticed, because we paid no attention to them. Only when we directed our attention to them, voluntarily or involuntarily, were impressions registered in the brain, their depth and durability being influenced by the amount of attention employed.

If the attention was already employed in a concentrated manner, for example reading a book, other sights and sounds would pass unnoticed. A person might try to speak to us and we might not hear them, until they attracted our attention in some way. Conversely, how often it happens that we read an entire page of a work, when thinking of something different from what we are reading, and without taking in anything of what we have completely read. Our attention had been distracted.

It is, then, *only noticed sensations*, that is, those which arrest attention, that give rise to ideas. ... Mammals have the same senses as man ... But, since they do not dwell on most of these sensations, nor fix their attention upon them, but only notice those that are immediately related to their usual needs, these animals have but a small number of ideas, which are always more or less the same with little or no variation ... The effects of education forced upon animals is well known.

... men who have not been compelled by an early education to exert their intellects ... are extremely limited as regards their moral wellbeing. The ideas which they form are almost entirely reduced to ideas of self-interest, property, and a few physical enjoyments ... education, which so wonderfully develops the human intellect, only achieves this result by imbuing a habit of thinking and fixing attention on ... numerous and varied objects ... Education thus inculcates a habit of exciting the intellect and varying the thoughts ... multiplying ideas of every kind, but especially complex ideas.

Darwin was also a great believer in the power of education but, whereas Lamarck seemed to have seen education as being of benefit to the individual, Darwin believed that the benefits would be passed on to the next generation by inheritance, bringing a far greater benefit to mankind. It was after Darwin that moves were made to introduce compulsory education for all children up to the age of 12, when, at that time, they could legally enter the workforce. Undoubtedly, compulsory education has wrought a great change in Western society. Has that change come about as the result of education's effect on multitudes of individuals, or does the extraordinary amount of technological advance which has been made during the past halfcentury indicate that Darwin was right, and that there is a cumulative effect over the generations?

The second faculty Lamarck discussed was thought.

Thought is the most universal of intellectual acts … Thought must be regarded as an action carried out in the organ of intelligence by movements of the nervous fluid. It works on ideas already acquired … as in memory; or by comparing some of these ideas together so as to draw judgments from them, or to ascertain their relations … as in reasoning; or by methodically dividing … them as in analysis; or, lastly, in creating new ideas … as in the operations of the imagination.

… thought which constitutes reflection … is more than an act of memory and yet is not a judgment.

Lamarck claimed that ideas were the raw material of intellectual operations and the nervous fluid was the sole agent which gave rise to them. The cerebral pulp being passive, the nervous fluid was the only active body. Beings endowed with intelligence had the capacity to guide the nervous fluid over the outlines impressed in the cerebral pulp made by a previously acquired idea. But if, instead of recalling a single idea, the individual recalls several, and carries out operations on these ideas, he then forms thoughts less simple and more prolonged … imagination … originates in the habit of thinking and forming complex ideas.

… thought is always accompanied by attention … when the latter ceases the former promptly comes to an end.

… since thought is an action, it uses up nervous fluid and, consequently, when it is maintained too long, it causes fatigue … if during [the] labour of the stomach you divert nervous fluid from the digestion towards the hypocephalon … you damage your digestion and endanger your health …

Lamarck recommended one task at a time. After eating, do not indulge in any activity which will result in the nervous fluid being diverted from its task of digestion. Early morning, after a good sleep, when the nervous fluid was very abundant, was the best time for intellectual or physical exercise. The more nervous energy an individual used for intellectual activity, the less that individual would have available for physical activity – and vice versa.

The third intellectual faculty considered by Lamarck was imagination.

The imagination is that faculty for creating new ideas that the organ of intelligence acquires by means of its thoughts. It is dependent upon the presence of many ideas, out of which new complex ideas are constantly being formed … Acts of the imagination consist in creating new ideas by comparisons and judgments of previous ideas … the individual can form … a number of new ideas … and out of these many more again …

207

Humans had the ability to imagine nebulous contrasts. Having experienced the finite, man could imagine the infinite; having experienced time, man could imagine infinity. No other animal could do this. Animals lacked imagination because they had few needs, varied their thoughts but little and had few ideas. Humans, living in such varied societies and under such varied conditions, had been forced to vary their ideas, form many complex ones and hence were possessed of the materials necessary for imagination. Like all other faculties, imagination increased the more it was used. Lamarck did not address the issue of the vivid imaginations possessed by some children, who would not have had the time to acquire the many and varied experiences which he deemed necessary to produce imagination.

Now genius in an individual is nothing else but a high imagination, guided by exquisite taste and a well-balanced judgment, and nurtured and enlightened by a vast knowledge, and controlled in short by a high degree of reason.

What would literature be without imagination! … How could poetry … dispense with imagination? … I hold that literature … is the noble and sublime art of arousing our passions, elevating and widening our thoughts … science is to that extent inferior; for she teaches coldly and stiffly; but literature is superior in this … that she also greatly broadens our thoughts by showing us everywhere what is really there and not what we want to find.

The purpose of the former is to give pleasure; that of the latter is to collect all practicable positive knowledge.

This being so, imagination is as much to be feared in the sciences as it is indispensable in literature … imagination nearly always gives rise to errors, when it is not controlled and limited by learning and reason ... yet without imagination there is no genius ... imagination is to be feared in the sciences ... when it is not controlled by a lofty and enlightened reason; when it is so controlled, it is one of the essential factors in the progress of science.

Reading the above, one cannot but wonder how many of the positive attributes there mentioned Lamarck believed that he, himself, possessed? Was he arrogant, or, at least, did his colleagues perceive him to be so? Was this the reason why his work was not embraced?

Despite my admiration for Lamarck, I felt uncomfortable reading this section, not because I thought that what Lamarck was saying was wrong, but because I felt he was placing himself in a superior position. Quite why pride in one's superior ability and achievements is acceptable in areas such as art, music, sport, literature or even mathematics, but not in the sphere of intellectual ability, is hard to explain. Nevertheless, what appears to be a claim by Lamarck that he was endowed with "lofty and enlightened reason" and the further implication that some of his colleagues, both past and present, were not, makes uncomfortable reading.

The organ of understanding was also the seat of the memory activated when nervous fluid moved over the areas of the brain in which ideas and sensations had previously been impressed. Memory may be described as the most important and necessary of the intellectual faculties, for without memory what could we do? ... Without memory, man would have no kind of knowledge ... he could not even have a language for the expression of his ideas ... he would be altogether deprived of the faculty of thought and imagination if he had no memory. ... memory can only come into existence after ideas have been acquired

… nature can have given to the most perfect animals and even to man nothing but memory; she cannot give prescience, that is to say, a knowledge of future events. In a footnote, Lamarck explained that it was possible for humans to know in advance the time of an eclipse, or the alignment of stars, but such certain knowledge of future events was very rare. Both humans and animals could anticipate other events, such as the change of the seasons, but not with precise knowledge of exactly what would happen when.

Memory occurred when the nervous fluid came into contact with images previously impressed upon the brain. The nervous fluid was controlled by the inner feeling which directed it. Persons, places, objects and events were also recalled by memory in the same way as ideas.

Lamarck then suggested that during perfect sleep, the nervous fluid was at rest; during imperfect sleep, when the nervous fluid was disturbed, perhaps by imperfect digestion, random ideas and thoughts were generated, but during sleep it was not possible to dream of anything not based on an image previously impressed on the brain, however disordered was the recall. When I see my dog dreaming, barking in his sleep, and giving unequivocal signs of the thoughts which agitate him, I become convinced that he too has ideas, of however limited a kind.

Lamarck then wrote of how the inner feeling of lunatics was unable to direct the movements of the nervous fluid into the hypocephalon. Lamarck seemed to perceive the problem to lie with the nervous fluid, rather than with the tissue itself. Delirium was also caused by a physical condition agitating and disordering the flow of the nervous fluid.

The last of the principal faculties of the intellect which Lamarck considered was judgment. Lamarck believed that in this faculty lay the origin of will, which gave rise to desires, wishes, hope, anxiety, fear, etc.

We cannot carry out any series of thoughts without forming judgments; our reasoning and analyses are pure results of judgments; our imagination itself has no power, except through its judgments …

Judgment, like every other faculty, needed to be practiced, which meant that each individual should observe the results of the judgments they made and learn from experience. Many people made few judgments for themselves but relied on the judgment of others. Lamarck believed that this, at least at times, stemmed from the practice of children being expected to believe and do what they were told, rather than thinking for themselves.

Judgments are always the result of a comparison of ideas previously acquired. Lamarck suggested that each idea engraven occupied a special site in the encephalon. The nervous fluid could traverse more than one site at a time, recombining to form a new, complex, idea or judgment. Wrong judgments may be the result of wrong information or information incorrectly evaluated, which may happen as the result of prejudice, upbringing, etc., as well as through a fault of the intellect.

Lamarck added a final section on reason, which he had not previously mentioned, because reason was not a faculty, but a condition of the intellect … reason is nothing more than a stage acquired in the rectitude of judgments. The new born child has no experience to guide him and makes mistakes, is deceived by his sense, with such things as distances, shapes, colours and consistency of objects. Reason helped rectify such errors of judgment … the extent of our reason is proportional to the rectitude of our judgment on all things. Any being with the intellectual capacity to make a judgment must possess some degree of reason. Humans have the greatest variety of clear ideas and the greatest degree of reason. However, animals also show an increase in their ability to reason in regard to the situations which they face as they grow older. Instinct is a compelling force aroused by the inner feeling without the intervention of thought and judgment. Animals may act from instinct or from reason.

Societies also had their own 'reasons' which underpinned their way of life. These differed from society to society in the same way that reason differed from person to person. Private 'reason' might differ from public 'reason'. Now both individual and public reason, when they find themselves exposed to any alteration, usually set up so great an obstacle to it, that it is often harder to secure the recognition of a truth than it is to discover it …

In spite of the errors into which I may have been led, the work may possibly contain ideas and arguments that will have a certain value for the advancement of knowledge, until such time as the great subjects, with which I have ventured to deal, are treated anew by men capable of shedding further light on them.

Epilogue

I AM amazed that Hugh Elliot took the time and trouble that he did to translate this epic work, which he quite clearly disdained.

I am even more amazed that, having translated it, word by word, Elliot still held it in disdain and appeared so to misunderstand it.

But my greatest amazement is reserved for those people, who, having read Lamarck's scholarly treatise, still hold it to be inferior to the confused, and confusing, theory offered by Darwin, with its plethora of 'perhaps', 'possibly' and 'probably's.

That living forms had changed over time was an idea put forward by Aristotle, as well as Lamarck's mentor, Buffon, and Charles Darwin's grandfather, Erasmus Darwin, but Lamarck was the first person to offer a comprehensive account of how evolution may have taken place. Lamarck is the true father of evolutionary theory.

It is my sincere hope that others may be inspired to read Lamarck's full work, to connect all the dots scattered throughout this book. Perhaps then Jean-Baptistes de Monet de Lamarck and his work will be elevated to the position of esteem they both so richly deserve.

Bibliography

Bateson, W. (1909) Heredity and Variation in Modern Lights. In
A.C. Seward (ed.), *Darwin and Modern Science*, pp.85-101.
Cambridge: Cambridge University Press

Blyth, E. (1835) An attempt to Classify the 'Varieties' of Animals.
(The Magazine of Natural History (London) 8: 40-53) In L.
Eiseley (1979) *Darwin and the Mysterious Mr. X.* London: J.M.
Dent and Ass.

Blyth, E. (1836) Observations on the Various Seasonal and Other
Changes which Regularly Take Place in Birds. (The Magazine
of Natural History (London), 9: 399) In L. Eiseley (1979)
Darwin and the Mysterious Mr. X. London: J.M. and Ass.

Blyth, E. (1837) On the Psychological Distinctions Between Man
and All Other Animals. (The Magazine of Natural History
(London), 1: 1-9, 77-85, 131-141) In L. Eiseley (1979) *Darwin
and the Mysterious Mr. X.* London: J.M. and Ass.

Buffon, G.L. (1781/1834) *Natural History, General and Particular.*
London: Thomas Kelly

Chambers, R. (1844/1994) *Vestiges of the Natural History of
Creation.* Chicago: University of Chicago Press

Dawkins, R. (1976) *The Selfish Gene.* Oxford: Oxford University
Press

Eldredge, N. and Gould, S.J. (1972) Punctuated Equilibrium: An
Alternative to Phyletic Gradualism. In T.J.M. Schopf (ed.),198

Models in Paleobiology, pp.82-225 San Francisco: Freeman, Cooper

Hollick, M. (2006) *The Science of Oneness*. Winchester, U.K: O Books

Honeywil, R. (2008) *Lamarck's Evolution: Two centuries of genius and jealousy*. Millers Point, N.S.W: Murdoch Books

Hutton, J. (1788/1794/1970) Systems of the Earth (1788), Observations on Granite (1794). In G.W. White (ed.) *Contributions to the History of Geology*. Vol.5. Facsimile reproduction. Darien, Conn: Hasner

Huxley, T.H. (1863/1959) *Man's Place in Nature*. Michigan: University of Michigan Press

Jung, C.G. (1939) *The Interpretation of the Personality*. New York: Farrar and Rinehart

Lamarck, J.B. (1809/1963) *Zoological Philosophy*. Translated by H. Elliot. New York: Hafner Publishing Co.

Lyell, C. (1830-1833/1997) *Principles of Geology*. (Vols.1-3) Abridged edition. (James A.) (2nd ed.) London: Penguin

Maupertuis, P.L-M. de (1753/1966) *The Earthly Venus*. Translated by S. Boas, New York: Johnson Reprint Corporation

Playfair, J. (1802/1956) *Illustrations of the Huttonian Theory of the Earth*. Edinburgh: William Creech. (Facsimile reproduction with Introduction by G.W. White.)

Sheldrake, R. (1988) The Presence of the Past. London: Fontana

Teilhard de Chardin, P. (1951/1955) *The Phenomenon of Man*. New York: HarperTorch Books

Teillhard de Chardin, P. (1956/1965) *The Appearance of Man*. London: Collins

Teilhard de Chardin, P. (1956/1966) Man's Place in Nature. London: Collins199

About the Author

Born in London, Denise Carrington-Smith came to Australia 1967, raising her family in Melbourne, where she lived for nearly thirty years. During that time, Denise took up the study and practice of yoga, which she taught for a number of years.

It was through yoga that Denise became interested in healing, training first as a natural therapist and homœopath and then as a psychologist and clinical hypnotherapist.

After lecturing for several years in herbalism and homœopathy, Denise became Principal of the Victorian College of Classical Homœopathy and also served as President of the Australian Federation of Homœopaths.

In 1995, Denise retired to Far North Queensland, returning to University where she took up the study of archæology, which study included theories of evolution, especially human evolution. These theories were the subject of her Doctoral thesis. It was during this time that Denise became interested in the work of Lamarck, the controversy which surrounded it, and its ultimate rejection, a rejection which she believes to have been politically motivated and totally unjustified.

Denise has seven children, seventeen grandchildren and four greatgrandchildren (so far).

www.ingramcontent.com/pod-product-compliance
Lightning Source LLC
Chambersburg PA
CBHW071556210326
41597CB00019B/3269